FUNDAMENTALS OF
MOLECULAR SPECTROSCOPY

FUNDAMENTALS OF MOLECULAR SPECTROSCOPY

by

C. N. BANWELL

Lecturer in Chemistry,
University of Sussex, Brighton, Sussex

McGRAW-HILL PUBLISHING COMPANY LIMITED
LONDON
New York · Toronto · Sydney

Published by
McGraw-Hill Publishing Company Limited
McGraw-Hill House, Maidenhead, Berkshire, England

94020

THIS BOOK HAS BEEN SET IN MONOPHOTO TIMES NEW ROMAN 10 ON 12 POINT
AND PRINTED AND BOUND IN GREAT BRITAIN BY
WILLIAM CLOWES AND SONS LIMITED, LONDON AND BECCLES

PREFACE

Spectroscopy in its various branches is becoming of ever-increasing value as an analytical tool, from the qualitative and quantitative analysis of mixtures, through the determination of the bond lengths and angles of an individual molecule, right down to the evaluation of the electron distribution within a complicated organic substance. This book is intended as a guide to the fundamentals of spectroscopy so that those who are able to profit from application of its techniques may at least know something of the language the spectroscopist uses. It is based in the main on supervisions given to Cambridge undergraduates and on a lecture course in molecular spectroscopy delivered to third year students at Sussex.

Each chapter after the first presents the fundamentals of a particular spectroscopic technique, and thus each is essentially self-contained. The book is intended, however, to be read as a whole, at least initially, since concepts introduced in early chapters are often used later without further discussion. In this way it is hoped that the essential unity of all forms of molecular spectroscopy will become apparent. Purely quantitative analytical applications of spectroscopy are not discussed.

Emphasis throughout the book is on physical principles rather than detailed mathematics; in particular the results of wave-mechanical calculations of energy levels and selection rules are stated where necessary but not derived. However, reference is made at the end of each chapter to the many excellent texts of a more exhaustive and mathematical nature in which the interested reader will find ample material to satisfy his curiosity in this respect.

I would like to thank those of my colleagues in this University who so helpfully criticized parts of the manuscript, and in particular Dr. J. G. Stamper for his comments on Chapters 5 and 6. I am especially indebted to Professor N. Sheppard of the University of East Anglia who, apart from being the mentor at whose feet I learned virtually all I know about spectroscopy, has painstakingly read the whole manuscript and made many valuable comments and suggestions which have, I feel sure, added greatly to the usefulness of this book.

C. N. Banwell

CONTENTS

CONTENTS

LIST OF TABLES

CHAPTER 1

INTRODUCTION

1. Characterization of Electromagnetic Radiation

Molecular spectroscopy may be defined as the study of the interaction of electromagnetic waves and matter. Throughout this
book we shall be concerned with what spectroscopy can tell us of
the structure of matter, so it is essential in this first chapter to discuss briefly the nature of electromagnetic radiation and the sort of
interactions which may occur; we shall also consider, in outline, the
experimental methods of spectroscopy.

Electromagnetic radiation, of which visible light forms an obvious
but very small part, may be considered as a simple harmonic wave
propagated from a source and travelling in straight lines except
when refracted or reflected. The properties which undulate—
corresponding to the physical displacement of a stretched string
vibrating, or the alternate compressions and rarefactions of the
atmosphere during the passage of a sound wave—are interconnected
electric and magnetic fields. We shall see later that it is these
undulatory fields which interact with matter giving rise to a spectrum.

It is trivial to show that any simple harmonic wave has properties
of the sine wave, defined by $y = A \sin \theta$, which is plotted in Fig. 1.1.

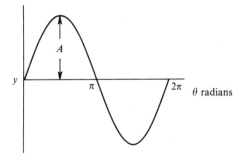

Fig. 1.1: The curve of $y = A \sin \theta$.

1

Here y is the displacement with a maximum value A, and θ is an angle varying between 0 and 360° (or 0 and 2π radians). The relevance of this representation to a travelling wave is best seen by considering the left-hand side of Fig. 1.2. A point P travels with uniform angular velocity ω rad/sec in a circular path of radius A; we measure the time from the instant when P passes O' and then after a time t sec, we imagine P to have described an angle $\theta = \omega t$ radians. Its vertical displacement is then $y = A \sin \theta = A \sin \omega t$,

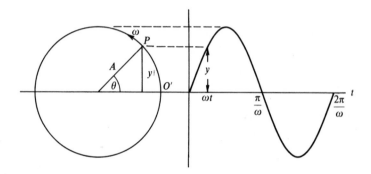

Fig. 1.2: The description of a sine curve in terms of the circular motion of a point P at a uniform angular velocity of ω radians/sec.

and we can plot this displacement against time as on the right-hand side of Fig. 1.2. After a time of $2\pi/\omega$ sec, P will return to O', completing a "cycle". Further cycles of P will repeat the pattern and we can describe the displacement as a continuous function of time by the graph of Fig. 1.2.

In one second the pattern will repeat itself $\omega/2\pi$ times, and this is referred to as the *frequency* (v) of the wave, measured in cycles per second (c/sec). We may then write

$$y = A \sin \omega t = A \sin 2\pi v t \qquad (1.1)$$

as a basic equation of wave motion.

So far we have discussed the variation of displacement with time, but in order to consider the nature of a *travelling* wave, we are more interested in the distance-variation of the displacement. For this we need the fundamental distance–time relationship:

$$x = ct, \qquad (1.2)$$

where x is the distance covered in time t at a speed c. Combining (1.1) and (1.2) we have:

$$y = A \sin 2\pi vt = A \sin \frac{2\pi vx}{c}$$

and the wave is shown in Fig. 1.3. Besides the frequency v, we now have another property by which we can characterize the wave—its *wave-length* λ, which is the distance travelled during a complete

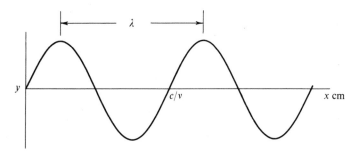

Fig. 1.3: The concept of a travelling wave with a wave-length λ.

cycle. When the velocity is c cm/sec and there are v c/sec, evidently there are v waves in c cm or

$$v\lambda = c; \qquad \lambda = c/v, \tag{1.3}$$

so we have:

$$y = A \sin \frac{2\pi x}{\lambda}. \tag{1.4}$$

In spectroscopy wave-lengths are expressed in a variety of units chosen so that in any particular range (see Fig. 1.4) the wave-length does not involve large powers of ten. Thus, in the microwave region, λ is measured in cm or mm, while in the infra-red it is usually given in microns (μ), where one micron is a millionth of a metre:

$$1 \, \mu = 10^{-6} \, \text{m} = 10^{-4} \, \text{cm}. \tag{1.5}$$

In the visible and ultra-violet region, λ is usually expressed in Ångstrom units (Å) where:

$$1 \, \text{Å} = 10^{-8} \, \text{cm} = 10^{-4} \, \mu. \tag{1.6}$$

There is yet another way in which electromagnetic radiation can be usefully characterized, and this is in terms of the *wavenumber* $\bar{v}$. Formally this is defined as the reciprocal of the wave-length expressed in centimetres:

$$\bar{v} = 1/\lambda \text{ cm}^{-1} \tag{1.7}$$

and hence

$$y = A \sin 2\pi\bar{v}x. \tag{1.8}$$

A more useful way of thinking of wavenumber, however, is as follows:

In one second radiation travels c cm and contains v (frequency) cycles. Hence each centimetre length contains v/c waves. But from equation (1.3),

$$v/c = 1/\lambda = \bar{v} \text{ cm}^{-1} \tag{1.9}$$

and we see that the wavenumber expresses the number of waves or cycles contained in each centimetre length of the radiation.

It is unfortunate that the symbols for wavenumber ($\bar{v}$) and frequency (v) are similar—and this particularly since some texts use v for wavenumber. Confusion should not arise, however, if the units of any expression are kept in mind, since wavenumber is in reciprocal centimetres (cm^{-1}) and frequency in c/sec. The two are, of course, proportional as we see from equation (1.9): $v = c\bar{v}$.

We thus have three properties by which we may characterize radiation—its frequency, its wave-length and its wavenumber. The three are easily interconvertible and all are extensively used by spectroscopists. The velocity of all electromagnetic radiation is a constant (very nearly 3×10^{10} cm/sec) when the radiation traverses a vacuum.

2. The Quantization of Energy

Towards the end of the last century experimental data were observed which were quite incompatible with the previously accepted view that matter could take up energy continuously. In 1900 Max Planck published the revolutionary idea that the energy of an oscillator is discontinuous and that any change in its energy content can occur only by means of a jump between two distinct energy states. The idea was later extended to cover many other forms of the energy of matter.

A molecule in space can have many sorts of energy; e.g. it may possess rotational energy by virtue of bodily rotation about its centre of gravity; it will have vibrational energy due to the periodic displacement of its atoms from their equilibrium positions; it will have electronic energy since the electrons associated with each atom or bond are in unceasing motion, etc. The chemist or physicist is early familiar with the electronic energy states of an atom or molecule and accepts the idea that an electron can exist in one of several discrete energy levels: he learns to speak of the energy as being *quantized*. In much the same way the rotational, vibrational and other energies of a molecule are also quantized—a particular molecule can exist in a variety of rotational, vibrational, etc., energy levels and can move from one level to another only by a sudden jump involving a finite amount of energy.

Consider two possible energy states of a system—two rotational energy levels of a molecule, for example—labelled E_1 and E_2 in the diagram. The suffixes 1 and 2 used to distinguish these levels are,

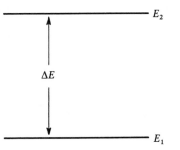

in fact, *quantum numbers*. The actual significance of quantum numbers goes far deeper than their use as a convenient label—in particular, we shall later see that analytical expressions for energy levels usually involve an algebraic function of one or more quantum numbers. Transitions can take place between the levels E_1 and E_2 provided the appropriate amount of energy, $\Delta E = E_2 - E_1$, can be either absorbed or emitted by the system. Planck suggested that such absorbed or emitted energy can take the form of electromagnetic radiation and that the frequency of the radiation has the simple form:

$$v = \Delta E/h \text{ c/sec}$$

i.e.

$$\Delta E = hv \text{ ergs} \tag{1.10}$$

where h is a universal constant—Planck's constant. This suggestion has been more than amply confirmed by experiment.

The significance of this is that if we take a molecule in state 1 and direct on to it a beam of radiation of a single frequency v (monochromatic radiation), where $v = \Delta E/h$, energy will be absorbed from the beam and the molecule will jump to state 2. A detector placed to collect the radiation after its interaction with the molecule will show that its intensity has decreased. Also if we use a beam containing a wide range of frequencies ("white" radiation), the detector will show that energy has been absorbed *only* from that frequency $v = \Delta E/h$, all other frequencies being undiminished in intensity. In this way we have produced a spectrum—an *absorption* spectrum.

Alternatively the molecule may already be in state 2 and may revert to state 1 with the consequent emission of radiation. A detector would show this radiation to have frequency $v = \Delta E/h$ only, and the *emission* spectrum so found is plainly complementary to the absorption spectrum of the previous paragraph.

The actual energy differences between the rotational, vibrational and electronic energy levels are very small and may be measured in ergs per molecule (or atom). In these units Planck's constant has the value:

$$h = 6 \cdot 62 \times 10^{-27} \text{ ergs sec molecule}^{-1}.$$

Often we are interested in the total energy involved when a gram-molecule of a substance changes its energy state: for this we multiply by the Avogadro number $N = 6 \cdot 02 \times 10^{23}$.

However, the spectroscopist measures the various characteristics of the absorbed or emitted radiation during transitions between energy states and he often, rather loosely, uses frequency, wavelength and wavenumber as if they were energy units. Thus in referring to "an energy of 10 cm^{-1}" he means: "a separation between two energy states such that the associated radiation has a wavenumber value of 10 cm^{-1}". The former expression is so simple and convenient that it is essential to become familiar with wavenumber and frequency energy units if one is to understand the spectroscopist's language. Throughout this book we shall use the symbol ε to represent energy in cm^{-1}.

It cannot be too firmly stressed at this point that the frequency of radiation associated with an energy change does *not* imply that the transition between energy levels occurs a certain number of times each second.   Thus an electronic transition in an atom or molecule may absorb or emit radiation of frequency some 10^{15} c/sec, but the electronic transition does not itself *occur* 10^{15} times per second.   It may occur once or many times and on each occurrence it will absorb or emit an energy quantum of the appropriate frequency.

3. Regions of the Spectrum

Figure 1.4 illustrates in pictorial fashion the various, rather arbitrary, regions into which electromagnetic radiation has been divided.   The boundaries between the regions are by no means precise, although the molecular processes associated with each region are quite different.   Each succeeding chapter in this book deals essentially with one of these processes.

In increasing frequency the regions are:

1. Radiofrequency region: 3×10^6–3×10^{10} c/sec;   10 m–1 cm wave-length.   Nuclear magnetic resonance (n.m.r.) and electron spin resonance (e.s.r.) spectroscopy.   The energy change involved is that arising from the reversal of spin of a nucleus or electron, and is of the order of 0·0001–1 cal/mole (Chapter 7).
2. Microwave region: 3×10^{10}–3×10^{12} c/sec; 1 cm–100 μ wave-length.   Rotational spectroscopy.   Separations between rotational levels are of the order of tens of cal/mole (Chapter 2).
3. Infra-red region: 3×10^{12}–3×10^{14} c/sec;   100 μ–1 μ wave-length.   Vibrational spectroscopy.   One of the most valuable spectroscopic regions for the chemist.   Separations between levels are some kilocalories per mole (Chapter 3).
4. Visible and ultra-violet regions: 3×10^{14}–3×10^{16} c/sec; 1 μ–100 Å wave-length.   Electronic spectroscopy.   The separations between the energies of valence electrons are some tens of kilocalories per mole (Chapters 5 and 6).
5. X-ray region: 3×10^{16} c/sec and above; 100 Å wave-length or less.   Energy changes involving the inner electrons of an atom or molecule which may be up to a thousand kilocalories.

X-ray spectroscopy is not discussed in this book since the data which it yields are of little importance for structural determinations. This is not to be confused with *X-ray diffraction* studies which can give a great deal of structural information; these, however, are not spectroscopic techniques, and fall outside the scope of this book.

2

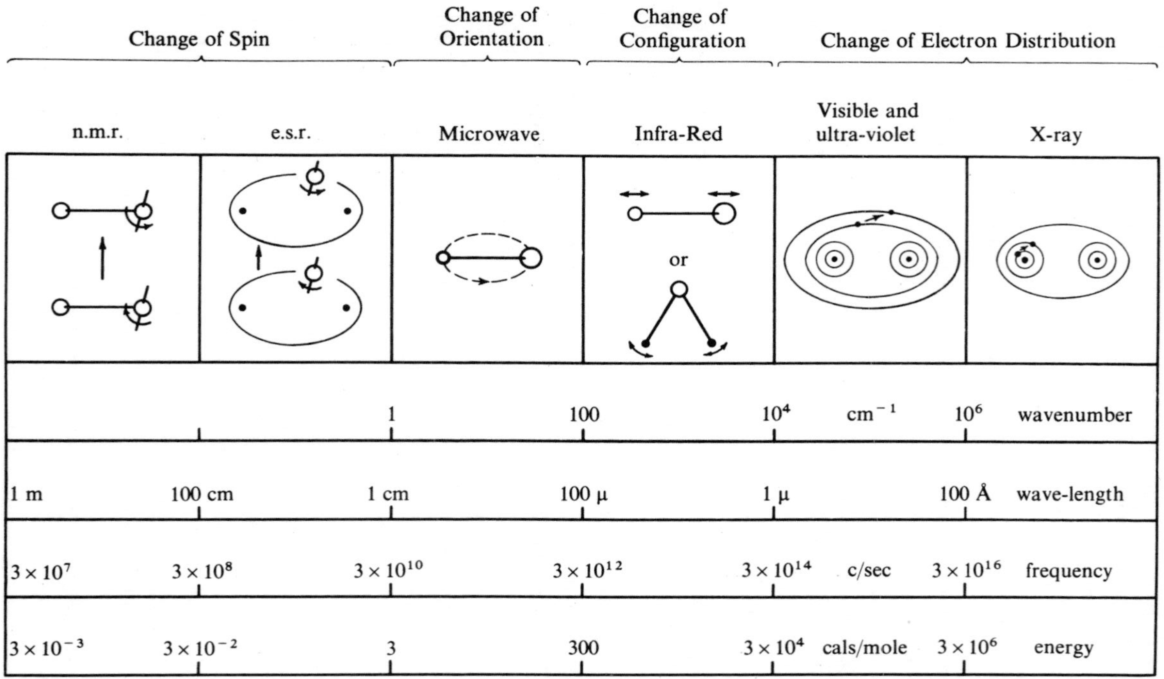

Change of Spin		Change of Orientation	Change of Configuration	Change of Electron Distribution	
n.m.r.	e.s.r.	Microwave	Infra-Red	Visible and ultra-violet	X-ray
1 m	100 cm	1	100	10^4	10^6	wavenumber
		cm^{-1}			
	1 cm	100 μ	1 μ	100 Å	wave-length
3×10^7	3×10^8	3×10^{10}	3×10^{12}	3×10^{14}	3×10^{16}	c/sec	frequency
3×10^{-3}	3×10^{-2}	3	300	3×10^4	3×10^6	cals/mole	energy

Fig. 1.4: *The regions of the electromagnetic spectrum.*

One other type of spectroscopy, that discovered by Raman and bearing his name, is discussed in Chapter 4. This, it will be seen, yields information similar to that obtained in the microwave and infra-red regions, although the experimental method is such that observations are made in the visible region.

In order that there shall be some mechanism for interaction between the incident radiation and the nuclear, molecular or electronic motions depicted in Fig. 1.4, there must be some electric or magnetic change produced by the motion which can be influenced by the electric or magnetic fields associated with the radiation. There are several possibilities:

(*i*) The radiofrequency region. We may consider the nucleus and electron to be tiny charged particles, and it follows that their spin is associated with a tiny magnetic dipole. The reversal of this dipole consequent upon the spin reversal can interact with the magnetic field of electromagnetic radiation at the appropriate frequency. Consequently all such spin reversals produce an absorption or emission spectrum.

(*ii*) The visible and ultra-violet region. The excitation of a valence electron involves the moving of electronic charges in the molecule. The consequent change in the electric dipole gives rise to a spectrum by its interaction with the undulatory electric field of radiation.

(*iii*) The microwave region. A molecule such as hydrogen chloride, HCl, in which one atom (the hydrogen) carries a permanent net positive charge and the other a net negative charge, is said to have a permanent electric dipole moment. H_2 or Cl_2, on the other hand, in which there is no such charge separation, have a zero dipole. If we consider the rotation of HCl (Fig. 1.5, where we notice that if only a pure rotation takes place, the centre of gravity of the molecule must not move), we see that the plus and minus charges change places periodically, and the component dipole moment in a given direction (say upwards in the plane of the paper) fluctuates regularly. This fluctuation is plotted in the lower half of Fig. 1.5, and it is seen to be exactly similar in form to the fluctuating electric field of radiation (cf. Fig. 1.2). Thus interaction can occur, energy can be absorbed or emitted, and the rotation gives rise to a spectrum. All molecules having a permanent moment are said to be "microwave active". If there is no dipole, as in H_2 or Cl_2, no interaction can take place and the molecule is "microwave in-active". This imposes a limitation on the applicability of micro-wave spectroscopy.

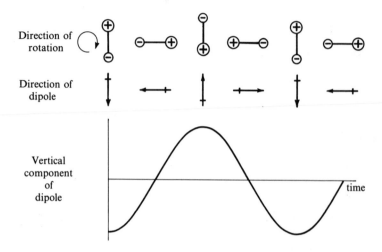

Fig. 1.5: *The rotation of a diatomic molecule, HCl, showing the fluctuation in the dipole moment measured in a particular direction.*

(*iv*) The infra-red region. Here it is a vibration, rather than a rotation, which must give rise to a dipole change. Consider the carbon dioxide molecule as an example, in which the three atoms are arranged linearly with a small net positive charge on the carbon and small negative charges on the oxygens:

$$\overset{\delta-}{O} \text{------} \overset{2\delta+}{C} \text{------} \overset{\delta-}{O}$$

During the mode of vibration known as the "symmetric stretch", the molecule is alternately stretched and compressed, both C—O bonds changing simultaneously, as in Fig. 1.6. Plainly the dipole

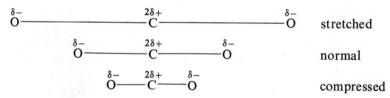

Fig. 1.6: *The symmetric stretching vibration of the carbon dioxide molecule.*

moment remains zero throughout the whole of this motion, and this particular vibration is thus "infra-red inactive".

However, there is another stretching vibration called the anti-symmetrical stretch, depicted in Fig. 1.7. Here one bond stretches while the other is compressed, and vice versa. As the figure shows,

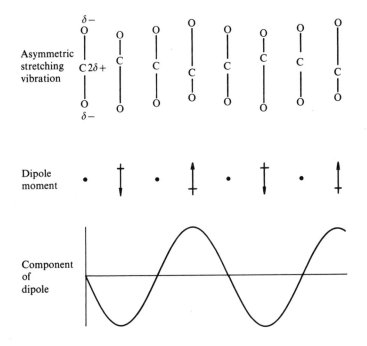

Fig. 1.7: The asymmetric stretching vibration of the carbon dioxide molecule showing the fluctuation in the dipole moment.

there is a periodic alteration in the dipole moment, and the vibration is thus "infra-red active". One further vibration is allowed to this molecule (see Chapter 3 for a more detailed discussion), known as the bending mode. This, as shown in Fig. 1.8, is also infra-red active. In both these motions the centre of gravity does not move.

Although dipole change requirements do impose some limitation on the application of infra-red spectroscopy, the appearance or non-appearance of certain vibration frequencies can give valuable information about the structure of a particular molecule (see Chapter 3).

(*v*) There is a rather special requirement for a molecular motion to be "Raman active"; this is that the electrical *polarizability* of the molecule must change during the motion. This will be discussed fully in Chapter 4.

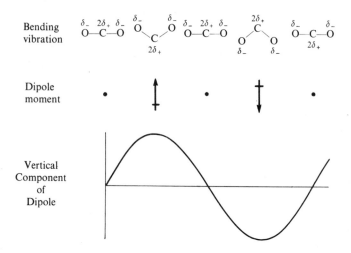

Fig. 1.8: The bending motion of the carbon dioxide molecule and its associated dipole fluctuation.

4. Representation of Spectra

We show in Fig. 1.9 a highly schematic diagram of a spectrometer suitable for use in the visible and ultra-violet regions of the spectrum. A "white" source is focused by lens (1) on to a narrow slit (arranged perpendicularly to the plane of the paper) and is then made into a parallel beam by lens (2). After passing through the sample it is separated into its constituent frequencies by a prism and is then focused on to a photographic plate by lens (3); the vertical image of the slit will thus appear on the plate. Rays have been drawn to show the points at which two frequencies, v_1 and v_2, are focused.

If the sample container is empty, the photographic plate, after development, should ideally show an even blackening over the whole range of frequencies covered (i.e. from A to B). The ideal situation is seldom realized, if only because the source does not

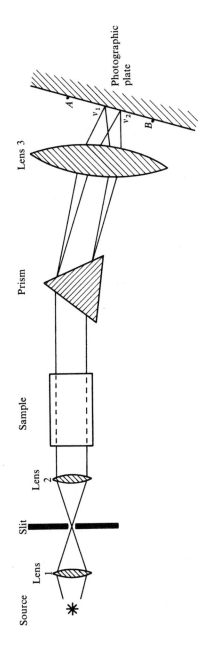

Fig. 1.9: Schematic diagram of a spectrometer suitable for operation in the visible region.

usually radiate all frequencies with the same intensity, but in any case the blackening of the plate serves to indicate the relative intensities of the frequencies emitted by the source.

If we now imagine the sample space to be filled with a substance having only two possible energy levels, E_1 and E_2, the photographic plate, after development, will show a blackening at all points except at the frequency $v = (E_2 - E_1)/h$, since energy at this frequency will have been absorbed by the sample in raising each molecule from state 1 to state 2. Further if, as is almost always the case, there are many possible energy levels, $E_1, E_2, \ldots, E_j, E_k, \ldots$ available to the sample, a series of absorption lines will appear on the photographic plate at frequencies given by $v = (E_j - E_k)/h$. A typical spectrum may then appear as in Fig. 1.10.

At this point it may be helpful to consider what happens to the energy absorbed in the sort of process described above. In the ultra-violet, visible and infra-red regions it is an experimental fact that a given sample continues to show an absorption spectrum for as long as we care to irradiate it—in other words, a finite number of sample molecules appears to be capable of absorbing an infinite amount of energy. Plainly the molecules must be able to rid themselves of the absorbed energy.

A possible mechanism for this is by thermal collisions. An energized molecule collides with its neighbours and gradually loses its excess energy to them as kinetic energy—the sample as a whole becomes warm.

Another mechanism is that energy gained from radiation is lost as radiation once more. A molecule in the ground state absorbs energy at frequency v and its energy is raised an amount $\Delta E = hv$ above the ground state. It is thus in an excited, unstable condition, but by emitting radiation of frequency v again, it can revert to the ground state and is able to re-absorb from the radiation beam once more. In this case, it is often asked how an absorption spectrum can arise at all, since the absorbed energy is re-emitted by the sample. The answer is simply that the radiation is re-emitted in a random direction and the proportion of such radiation reaching the detector is minute—in fact re-emitted radiation has as much chance of reaching the source as the detector! The net effect, then, is an absorption from the directed beam and, when re-emission occurs, a scattering into the surroundings. The scattered radiation can, of course, be collected and observed as an emission spectrum which

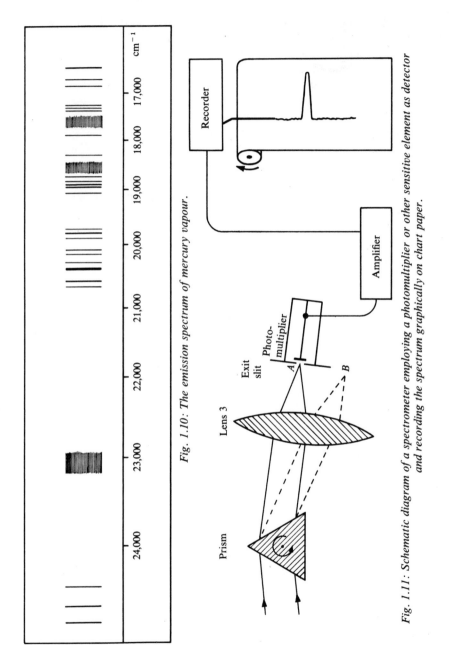

Fig. 1.10: The emission spectrum of mercury vapour.

Fig. 1.11: Schematic diagram of a spectrometer employing a photomultiplier or other sensitive element as detector and recording the spectrum graphically on chart paper.

will be—with important reservations to be discussed in Chapter 4—
the complement of the absorption spectrum.

In modern spectrometers the detector is rarely the simple photographic plate of Fig. 1.9.   One of the most sensitive and useful
devices in the visible and ultra-violet region is the photomultiplier
tube, consisting of a light-sensitive surface which emits electrons
when light falls upon it.   The tiny electron current may be amplified and applied to an ammeter or pen recorder.   The spectrometer would then appear somewhat as in Fig. 1.11, where the
sensitive element of the photomultiplier is situated at the point A of
Fig. 1.9.   The physical width of the beam falling on the detector
can be limited by the provision of an "exit slit" just in front of the
detector entrance.

The frequency of the light falling on the photomultiplier may be
altered either by physically moving the latter from A to B or, more
usually, by steady rotation of the prism.   If, as before, we imagine
the sample to contain a substance having just two energy levels, the
photomultiplier output will, ideally, vary with the prism orientation
as in Fig. 1.12.   We say that the spectrum has been *scanned* between the frequencies represented by A and B.

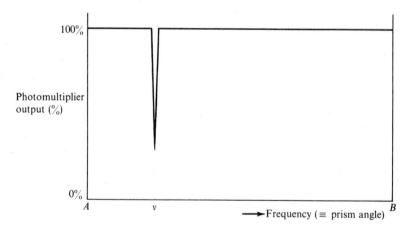

Fig. 1.12: The idealized spectrum of a molecule undergoing a single transition.

Again, the ideal situation is seldom attained.   Not only does the
source emissivity vary with frequency, but often the sensitivity of
the photomultiplier is also frequency dependent.   Thus the base-
line—the "sample-empty" condition—is never horizontal, although

matters can usually be arranged so that it is approximately linear. Further, since it is impossible to make either of the slits infinitely narrow, a *range* of frequencies, rather than just a single frequency, falls on the photomultiplier at any given setting of the prism.   This results in a broadening of the absorbance peak, and the final spectrum may appear rather as in Fig. 1.13.   In this figure, too, we have plotted absorbance upwards from 0 to 100% and transmittance—its complement—downwards.   This is the usual way in which such spectra are represented.

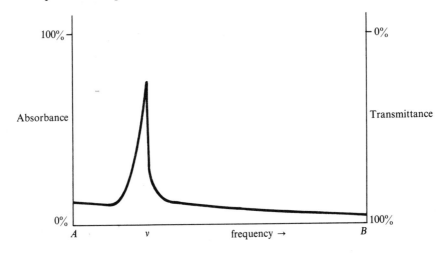

Fig. 1.13: *The usual appearance of the spectrum of a molecule undergoing a single transition (cf. Fig. 1.12); here the background is no longer constant and the absorption region is of finite width.*

If there are again several energy levels available to the sample, it is very unlikely that there is the same probability of transition between each of them.   The question of transition probability will be raised again more formally in each chapter, but here we may note that differences in transition probability will mean that the absorbance (or transmittance) at each absorbing frequency will differ. This is shown by the varying intensities of the lines on the photographic plate of Fig. 1.10, and, more precisely, by the recorder trace of Fig. 1.14(a).

Figure 1.14(a) shows the sort of record which is produced by most modern spectrometers, whatever the region in which they

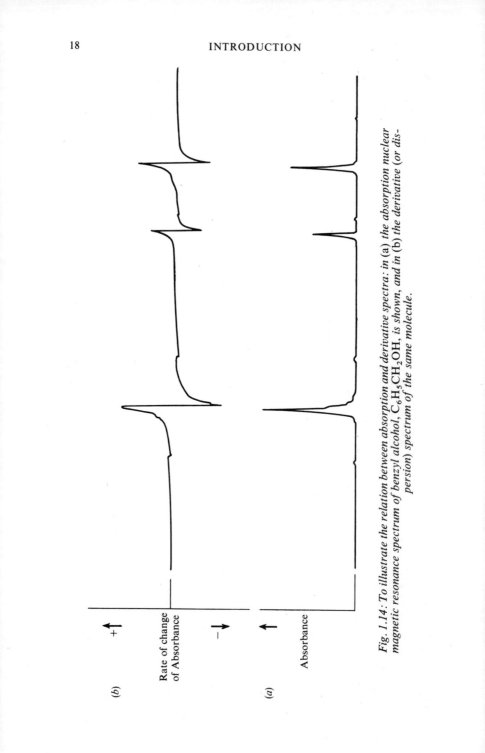

Fig. 1.14: To illustrate the relation between absorption and derivative spectra: in (a) *the absorption nuclear magnetic resonance spectrum of benzyl alcohol,* $C_6H_5CH_2OH$, *is shown, and in* (b) *the derivative (or dispersion) spectrum of the same molecule.*

operate. One other form of presentation is often adopted, how-ever, particularly in the microwave and radiofrequency regions, and this is to record the *derivative* of the spectral trace instead of the trace itself. The derivative of a curve is simply its slope at a given point. In calculus notation, the derivative of the spectral trace is dT/dv, where T is the transmittance. The derivative record is thus a plot of the slope dT/dv against v; this is shown in Fig. 1.14(*b*) corresponding to the T/v plot of Fig. 1.14(*a*).

Although at first sight more complex, the derivative trace has advantages over the direct record in some circumstances. Firstly, it indicates rather more precisely the centre of each absorbance peak: at the centre of a peak, the T curve is horizontal, hence dT/dv is zero, and the centres are marked by the intersection of the deriva-tive curve with the axis. Further, for instrumental reasons, it is often better to measure the relative intensities of absorbance peaks from the derivative curve than from the direct trace.

5. Basic Elements of Practical Spectroscopy

Spectrometers used in various regions of the spectrum naturally differ widely from each other in construction. These differences will be discussed in more detail in the following chapters, but here it will probably be helpful to indicate the basic features which are common to all types of spectrometer. We may, for this purpose, consider absorption and emission spectrometers separately.

(*i*) *Absorption Instruments.* Figure 1.15(*a*) shows, in block dia-gram form, the components of an absorption spectrometer which might be used in the infra-red, visible and ultra-violet regions. The radiation from a white source is directed by some guiding device (e.g. the lens of Fig. 1.9, or mirrors) on to the sample, from which it passes through an analyser (e.g. the prism of Fig. 1.9), which selects the frequency reaching the detector at any given time. The signal from the latter passes to a recorder which is synchronized with the analyser so as to produce a trace of the absorbance as the frequency varies.

In the microwave and radiofrequency region it is possible to construct monochromatic sources whose emission frequency can be varied over a range. In this case, as Fig. 1.15(*b*) shows, no analyser is necessary, the source being, in a sense, its own analyser. Now, it is necessary for the recorder to be synchronized with the source-scanning device in order that a spectral trace be obtained.

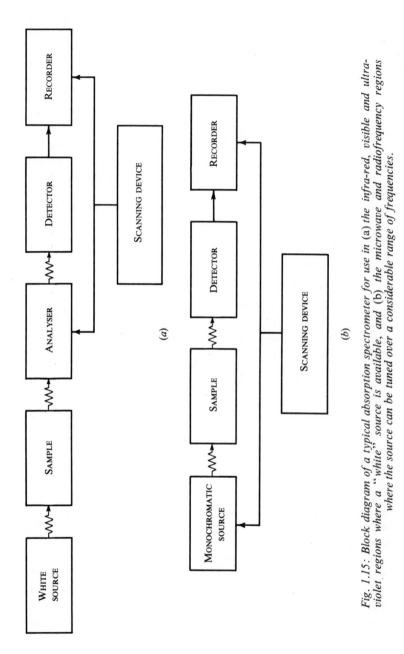

Fig. 1.15: Block diagram of a typical absorption spectrometer for use in (a) the infra-red, visible and ultra-violet regions where a "white" source is available, and (b) the microwave and radiofrequency regions where the source can be tuned over a considerable range of frequencies.

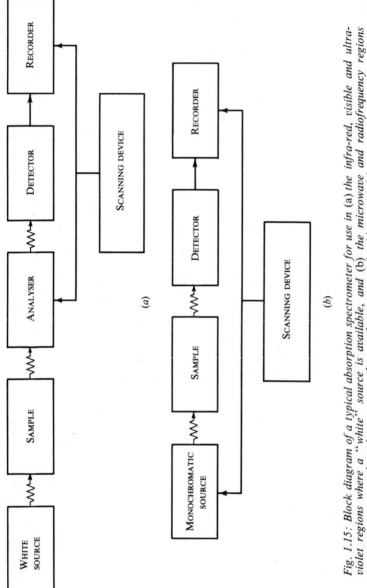

(*ii*) *Emission Instruments.* The lay-out now differs in that the sample, after excitation, is its own source, and it is necessary only to collect the emitted radiation, analyse and record it in the usual way. Figure 1.16 shows, schematically, a typical spectrometer. The

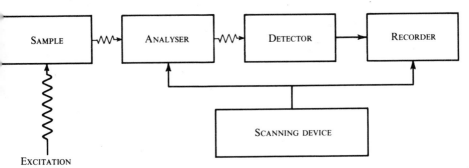

Fig. 1.16: Block diagram of a typical emission spectrometer.

excitation can be thermal or electrical, but often takes the form of electromagnetic radiation. In the latter case it is essential that the detector does not collect radiation directly from the exciting beam, and the two are usually placed at right angles as shown.

6. Signal-to-Noise: Resolving Power

Two other spectroscopic terms may be conveniently discussed at this point since they will recur in succeeding chapters.

(*i*) *Signal-to-Noise Ratio.* Since almost all modern spectrometers use some form of electronic amplification to magnify the signal produced by the detector, every recorded spectrum has a background of random fluctuations caused by spurious electronic signals produced by the detector, or generated in the amplifying equipment. These fluctuations are usually referred to as "noise". In order that a real spectral peak should show itself as such and be sufficiently distinguished from the noise, it must have an intensity some three or four times that of the noise fluctuations (a signal-to-noise ratio of three or four). This requirement places a lower limit on the intensity of observable signals.

(*ii*) *Resolving Power.* This is a somewhat imprecise concept which can, however, be defined rather arbitrarily and is often used as a measure of the performance of a spectrometer. We shall here consider it in general terms only.

No molecular absorption takes place at a single frequency only, but always over a spread of frequencies, usually very narrow but sometimes quite large. Such spreading can usually be attributed to molecular motions and interactions; it is for this reason that we have up to now drawn spectra with broadened line shapes (cf. Fig. 1.14(*a*)).

Let us consider two such lines close together, as on the right of Fig. 1.17(*a*): the dotted curve represents the absorption due to each line separately, the full line their combined absorption. We shall

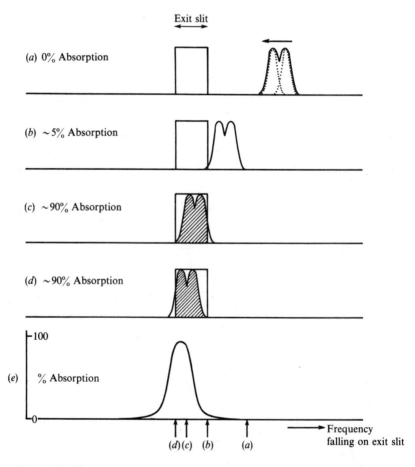

Fig. 1.17: *Illustrating the relation between slit width and resolving power:*
see text for discussion.

first take the exit slit width to be larger than the separation between the lines.   Scanning the spectrum plainly involves moving the twin absorbance peaks steadily to the left so that they pass across the exit slit and into the detector; the situation at successive stages is shown in (b), (c) and (d) of Fig. 1.17, the shaded area showing the amount of absorbance which the detector would register.   At (e) of this figure, the absorbance is plotted against frequency, together with the approximate positions of stages (a), (b), (c) and (d).

It is quite evident that the separation between the lines has disappeared under these conditions—the lines are not *resolved*.   It is equally evident that the use of a much narrower slit would result in their resolution—the *resolving power* would be increased.   In fact, provided the slit width is less than the separation between the lines, the detector output will show a minimum between them.

However, it must be remembered that a narrower slit allows less total energy from the beam to reach the detector and consequently the intrinsic signal strength will be less.   There comes a point when decreasing the slit width results in such weak signals that they become indistinguishable from the background noise mentioned in the previous paragraph.   Thus spectroscopy is a continual battle to find the minimum slit width consistent with acceptable signal-to-noise values.   Improvements in resolving power may arise not only as a result of obtaining better dispersion of the radiation by the analyser (e.g. by the use of a diffraction grating rather than a prism for the ultra-violet and infra-red regions) but also by using a more sensitive detector.

3

CHAPTER 2

MICROWAVE SPECTROSCOPY

1. The Rotation of Molecules

We saw in the previous chapter that spectroscopy in the microwave region is concerned with the study of rotating molecules. The rotation of a three-dimensional body may be quite complex and it is convenient to resolve it into rotational components about three mutually perpendicular directions through the centre of gravity—the principal axes of rotation. Thus a body has three principal *moments of inertia*, one about each axis, usually designated I_A, I_B and I_C.

Molecules may be classified into groups according to the relative values of their three principal moments of inertia—which, it will be seen, is tantamount to classifying them according to their shapes. We shall describe this classification here before discussing the details of the rotational spectra arising from each group.

(*i*) *Linear Molecules.* These, as the name implies, are molecules in which all the atoms are arranged in a straight line, such as hydrogen chloride HCl, or carbon oxy-sulphide OCS, illustrated below. The three directions of rotation may be taken as (*a*) about

$$H\text{——}Cl$$
$$O\text{——}C\text{——}S$$

the bond axis, (*b*) end-over-end rotation in the plane of the paper, and (*c*) end-over-end rotation at right angles to the plane. It is self-evident that the moments of (*b*) and (*c*) are the same (i.e. $I_B = I_C$) while that of (*a*) is very small. As an approximation we may say that $I_A = 0$, although it should be noted that this *is* only an approximation (see p. 31).

Thus for linear molecules we have:

$$I_B = I_C; \qquad I_A = 0. \tag{2.1}$$

(*ii*) *Symmetric Tops.* Consider a molecule such as methyl

fluoride, where the three hydrogen atoms are bonded tetrahedrally to the carbon, as shown below. As in the case of linear molecules,

the end-over-end rotation in, and out of, the plane of the paper are still identical and we have $I_B = I_C$. The moment of inertia about the C—F bond axis (chosen as the main rotational axis since the centre of gravity lies along it) is now not negligible, however, because it involves the rotation of three comparatively massive hydrogen atoms off this axis. Such a molecule spinning about this axis can be imagined as a top, and hence the name of the class. We have then:

$$\text{Symmetric tops: } I_B = I_C \neq I_A; \qquad I_A \neq 0. \qquad (2.2)$$

There are two subdivisions of this class which we may mention: if, as in methyl fluoride above, $I_B = I_C > I_A$, then the molecule is called a *prolate* symmetric top; whereas if $I_B = I_C < I_A$, it is referred to as *oblate*. An example of the latter type is boron trichloride, which, as shown, is planar and symmetrical. In this case:

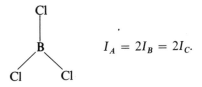

$$I_A = 2I_B = 2I_C.$$

(iii) *Spherical Tops.* When a molecule has all three moments of inertia identical, it is called a spherical top. A simple example is the tetrahedral molecule methane CH_4. We have then:

$$\text{Spherical tops: } I_A = I_B = I_C. \qquad (2.3)$$

In fact these molecules are only of academic interest in this chapter. Since they can have no dipole moment owing to their symmetry, rotation alone can produce no dipole change and hence no rotational spectrum is observable.

(*iv*) *Asymmetric Tops*. These molecules, to which the majority of substances belong, have all three moments of inertia different:

$$I_A \neq I_B \neq I_C. \tag{2.4}$$

Simple examples are water H_2O, and vinyl chloride $CH_2{=}CHCl$.

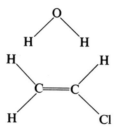

Perhaps it should be pointed out that one can (and often does) describe the classification of molecules into the four rotational classes in far more rigorous terms than have been used above (see e.g. Herzberg, *Molecular Spectra and Molecular Structure*, Vol. II). However, for the purposes of this book the above description is adequate.

2. Rotational Spectra

We have seen that rotational energy, along with all other forms of molecular energy, is quantized: this means that a molecule cannot have any arbitrary amount of rotational energy (i.e. any arbitrary value of angular momentum) but its energy is limited to certain definite values depending on the shape and size of the molecule concerned. The permitted energy values—the so-called rotational energy *levels*—may in principle be calculated for any molecule by solving the Schrödinger equation for the system represented by that molecule. For simple molecules the mathematics involved is straightforward but tedious, while for complicated systems it is probably impossible without gross approximations. We shall not concern ourselves unduly with this, however, being content merely to accept the results of existing solutions and to point out where reasonable approximations may lead.

We shall consider each class of rotating molecule in turn, discussing the linear molecule in most detail, because much of its treatment can be directly extended to symmetrical and unsymmetrical molecules.

3. Diatomic Molecules

3.1. The Rigid Diatomic Molecule. We start with this, the simplest of all linear molecules, shown in Fig. 2.1. Masses m_1 and m_2 are joined by a rigid bar (the bond) whose length is

$$r_0 = r_1 + r_2. \tag{2.5}$$

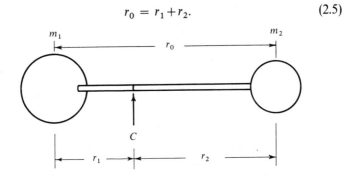

Fig. 2.1: *A rigid diatomic molecule treated as two masses, m_1 and m_2, joined by a rigid bar of length $r_0 = r_1 + r_2$.*

The molecule rotates end-over-end about a point C, the centre of gravity: this is defined by the moment, or balancing, equation:

$$m_1 r_1 = m_2 r_2. \tag{2.6}$$

The moment of inertia about C is defined by:

$$
\begin{aligned}
I &= m_1 r_1^2 + m_2 r_2^2 \\
&= m_2 r_2 r_1 + m_1 r_1 r_2 \quad \text{(from (2.6))} \\
&= r_1 r_2 (m_1 + m_2). \tag{2.7}
\end{aligned}
$$

But, from (2.5) and (2.6):

$$m_1 r_1 = m_2 r_2 = m_2 (r_0 - r_1)$$

therefore,

$$r_1 = \frac{m_2 r_0}{m_1 + m_2} \quad \text{and} \quad r_2 = \frac{m_1 r_0}{m_1 + m_2}. \tag{2.8}$$

Replacing (2.8) into (2.7):

$$I = \frac{m_1 m_2}{m_1 + m_2} r_0^2 = \mu r_0^2 \qquad (2.9)$$

where we have written $\mu = m_1 m_2/(m_1 + m_2)$, and μ is called the *reduced mass* of the system. Equation (2.9) defines the moment of inertia conveniently in terms of the atomic masses and the bond length.

By the use of the Schrödinger equation it may be shown that the rotational energy levels allowed to the rigid diatomic molecule are given by the expression:

$$E_J = \frac{h^2}{8\pi^2 I} J(J+1) \text{ ergs}, \quad \text{where } J = 0, 1, 2, \dots \qquad (2.10)$$

In this expression h is Planck's constant, and I is the moment of inertia, either I_B or I_C, since both are equal. The quantity J, which can take integral values from zero upwards, is called the *rotational quantum number*: its restriction to integral values arises directly out of the solution to the Schrödinger equation and is by no means arbitrary, and it is this restriction which effectively allows only certain discrete rotational energy levels to the molecule.

Equation (2.10) expresses the allowed energies in ergs; we, however, are interested in differences between these energies, or, more particularly, in the corresponding frequency, $v = \Delta E/h$ c/sec, or wavenumber, $\bar{v} = \Delta E/hc$ cm^{-1}, of the radiation emitted or absorbed as a consequence of changes between energy levels. In the rotational region spectra are usually discussed in terms of wavenumber, so it is useful to consider energies expressed in these units. We write:

$$\varepsilon_J = \frac{E_J}{hc} = \frac{h}{8\pi^2 Ic} J(J+1) \text{ cm}^{-1}, \quad J = 0, 1, 2, \dots \qquad (2.11)$$

where we shall consistently use the symbol ε to represent energy/ molecule in cm^{-1}.

Equation (2.11) is usually abbreviated to:

$$\varepsilon_J = BJ(J+1) \text{ cm}^{-1}, \quad J = 0, 1, 2, \dots \qquad (2.12)$$

where B, the *rotational constant*, is given by

$$B = \frac{h}{8\pi^2 I_B c} \text{ cm}^{-1} \qquad (2.13)$$

in which we have used explicitly the moment of inertia I_B. We might equally well have used I_C and a rotational constant C, but the notation of (2.13) is conventional.

From equation (2.12) we can show the allowed energy levels diagrammatically as in Fig. 2.2. Plainly for $J=0$ we have $\varepsilon_J=0$ and we would say that the molecule is not rotating at all. For $J=1$, the rotational energy is $\varepsilon_1=2B$ and a rotating molecule then has its lowest angular momentum. We may continue to calculate ε_J with increasing J values and, in principle, there is no limit to the rotational energy the molecule may have. In practice, of course, there comes a point at which the centrifugal force of a rapidly rotating diatomic molecule is greater than the strength of the bond, and the molecule is disrupted, but this point is not reached at normal temperatures.

We now need to consider *differences* between the levels in order to discuss the spectrum. If we imagine the molecule to be in the $J=0$ state (the *ground rotational state*, in which no rotation occurs), we can let incident radiation be absorbed to raise it to the $J=1$ state. Plainly the energy absorbed will be:

$$\varepsilon_{J=1} - \varepsilon_{J=0} = 2B - 0 = 2B \text{ cm}^{-1}$$

and, therefore,

$$\bar{v}_{J=0 \to J=1} = 2B \text{ cm}^{-1}. \tag{2.14}$$

In other words, an absorption line will appear at $2B$ cm^{-1}. If now the molecule is raised from the $J=1$ to the $J=2$ level by the absorption of more energy, we see immediately:

$$\bar{v}_{J=1 \to J=2} = \varepsilon_{J=2} - \varepsilon_{J=1}$$
$$= 6B - 2B = 4B \text{ cm}^{-1}. \tag{2.15}$$

In general, to raise the molecule from the state J to state $J+1$, we would have:

$$\bar{v}_{J \to J+1} = B(J+1)(J+2) - BJ(J+1)$$
$$= B[J^2 + 3J + 2 - (J^2 + J)]$$

or

$$\bar{v}_{J \to J+1} = 2B(J+1) \text{ cm}^{-1}. \tag{2.16}$$

Thus a step-wise raising of the rotational energy results in an absorption spectrum consisting of lines at $2B$, $4B$, $6B, \ldots$, cm^{-1}, while a similar lowering would result in an identical emission spectrum.   This is shown at the foot of Fig. 2.3.

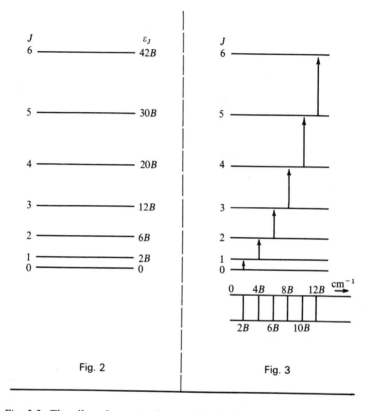

Fig. 2

Fig. 3

Fig. 2.2: The allowed rotational energy levels of a rigid diatomic molecule.

Fig. 2.3: Allowed transitions between the energy levels of a rigid diatomic molecule and the spectrum which arises from them.

In deriving this pattern we have made the assumption that a transition can occur from a particular level only to its immediate neighbour, either above or below: we have not, for instance, considered the sequence of transitions $J=0 \rightarrow J=2 \rightarrow J=4 \ldots$.   In fact, a rather sophisticated application of the Schrödinger wave equation shows that, for this molecule, we need only consider transitions in which J changes by one unit—all other transitions being spectro-scopically *forbidden*.   Such a result is called a *selection rule*, and we may formulate it for the rigid diatomic rotator as:

$$\text{Selection rule;} \quad \Delta J = \pm 1. \tag{2.17}$$

Thus equation (2.16) gives the *whole* spectrum to be expected from such a molecule.

Of course, only if the molecule is asymmetric (heteronuclear) will this spectrum be observed, since if it is homonuclear there will be no dipole component change during the rotation, and hence no inter-action with radiation.   Thus molecules such as HCl and CO will show a rotational spectrum, while N_2 and O_2 will not.   Remember, also, that rotation about the bond axis was rejected in Section 1: we can now see that there are two reasons for this.   Firstly, the moment of inertia is very small about the bond so, applying equa-tions (2.10) or (2.11) we see that the energy levels would be extremely widely spaced: this means that a molecule requires a great deal of energy to be raised from the $J=0$ to the $J=1$ state and such transi-tions do not occur under normal spectroscopic conditions.   Thus diatomic (and all linear) molecules are in the $J=0$ state for rotation about the bond axis, and they may be said to be not rotating. Secondly, even if such a transition should occur, there will be no dipole change and hence no spectrum.

To conclude this section we shall apply equation (2.16) to an observed spectrum in order to determine the moment of inertia and hence the bond length.   Gilliam *et al.** have measured the first line $(J=0)$ in the rotation spectrum of carbon monoxide as $3\cdot84235$ cm^{-1}.   Hence from equation (2.16):

$$\bar{v}_{0\rightarrow1} = 3\cdot84235 = 2B \text{ cm}^{-1},$$

or,

$$B = 1\cdot92118 \text{ cm}^{-1}.$$

Rewriting equation (2.13) as: $I = h/8\pi^2 Bc$, we have,

$$I_{CO} = \frac{6\cdot624 \times 10^{-27}}{8\pi^2 \times 2\cdot99776 \times 10^{10} \times B} = \frac{27\cdot9865 \times 10^{-40}}{B}$$

$$= 1\cdot45673_5 \times 10^{-39} \text{ g cm}^2.$$

---

* Gilliam, Johnson and Gordy, *Physical Review*, **78**, 140 (1950).

But the moment of inertia is μr^2 (cf. equation (2.9)) and, taking the atomic masses as $C = 12$, $O = 15\cdot9949$, we have

$$\mu = \frac{12 \times 15\cdot9949}{27\cdot9949 \times 6\cdot0244 \times 10^{23}} \text{ g/molecule.}$$

Hence:

$$r^2 = \frac{I}{\mu} = 1\cdot2800 \times 10^{-16} \text{ cm}^2$$

or

$$r_{CO} = 1\cdot131 \text{ Å.}$$

3.2. The Intensities of Spectral Lines.   We want now to consider briefly the relative intensities of the spectral lines of equation (2.16); for this a prime requirement is plainly a knowledge of the relative probabilities of transition between the various energy levels.   Does, for instance, a molecule have more or less chance of making the transition $J = 0 \rightarrow J = 1$ than the transition $J = 1 \rightarrow J = 2$?   We mentioned above calculations which show that a change of $\Delta J = \pm 2$, ± 3, etc., was forbidden—in other words, the transition probability for all these changes is zero.   Precisely similar calculations show that the probability of all changes with $\Delta J = \pm 1$ is the same—all are equally likely to occur.

This does not mean, however, that all spectral lines will be equally intense.   Although the intrinsic probability that a single molecule in the $J = 0$ state, say, will move to $J = 1$ is the same as that of a single molecule moving from $J = 1$ to $J = 2$, in an assemblage of molecules, such as in a normal gas sample, there will be different numbers of molecules in each level to begin with, and therefore different total numbers of molecules will carry out transitions between the various levels.   In fact, since the intrinsic probabilities are identical, the line intensities will be directly proportional to the initial numbers of molecules in each level.

The first factor governing the population of the levels is the Boltzmann distribution.   This statistical law states that in a system of N total molecules, a fraction N_J/N will occupy a particular energy level E_J where:

$$\frac{N_J}{N} = \frac{\exp{(-E_J/kT)}}{f}, \qquad (2.18)$$

where N_J=number of molecules in level J, k=Boltzmann's constant, T=temperature in °Abs. and f is a proportionality constant, sometimes called the partition function. Since the sum of all fractions N_J/N must equal unity, we have

$$\sum \frac{N_J}{N} = 1 = \frac{\sum \exp(-E_J/kT)}{f}$$

or

$$f = \sum \exp(-E_J/kT).$$

We are, however, interested only in *relative* populations of the energy levels given by equation (2.12), and may write:

$$N_J \propto \exp(-E_J/kT) \propto \exp(-BhcJ(J+1)/kT) \qquad (2.19)$$

as representing the population of each level. Plainly N_J decreases rapidly with increasing J, as shown in Fig. 2.4.

A second factor is also required—the possibility of *degeneracy* in the energy states. Degeneracy is the existence of two or more energy states which have exactly the *same* energy. In the case of the diatomic rotator we may approach the problem in terms of its angular momentum.

The defining equations for the energy and angular momentum of a rotator are:

$$E = \tfrac{1}{2}I\omega^2, \qquad \mathbf{P} = I\omega,$$

where I is the moment of inertia, ω the rotational frequency (in radians per sec), and $\mathbf{P}$ the angular momentum. Rearrangement of these gives

$$\mathbf{P} = \sqrt{2EI}.$$

The energy level expression of equation (2.10) can be rewritten:

$$2EI = J(J+1)\frac{h^2}{4\pi^2}$$

and hence

$$\mathbf{P} = \sqrt{J(J+1)}\,\frac{h}{2\pi} = \sqrt{J(J+1)} \text{ a.m.u.} \qquad (2.20)$$

where, following convention, we take $h/2\pi$ as the fundamental unit of angular momentum (a.m.u.). Thus we see that **P**, like *E*, is quantized.

Throughout the above derivation **P** has been printed in bold face type to show that it is a *vector*—i.e. it has *direction* as well as *magnitude*. The direction of the angular momentum vector is conventionally taken to be along the axis about which rotation occurs

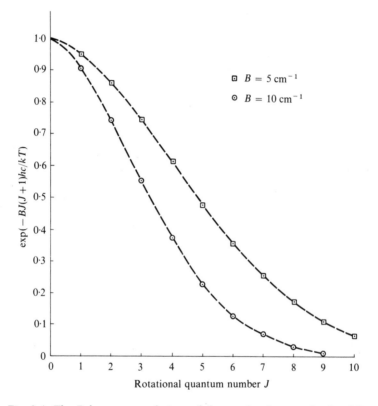

Fig. 2.4: *The Boltzmann populations of the rotational energy levels of Fig. 2.2: the diagram has been drawn taking values of B=5 and 10 cm^{-1} and T=300° Abs. in equation (2.19).*

and it is usually drawn as an arrow of length proportional to the magnitude of the momentum. The number of different directions which an angular momentum vector may take up is limited by a quantum mechanical law which may be stated:

"For integral values of the rotational quantum number (in this case J), the angular momentum vector may only take up directions such that its component along a given reference direction is zero or an integral multiple of a.m. units."

We can see the implications of this most easily by means of a diagram. In Fig. 2.5 we show the case $J=1$. Here $\mathbf{P}=\sqrt{1\times2}$ a.m.u. $=\sqrt{2}$, and, as Fig. 2.5(a) shows, a vector of length $\sqrt{2}\,(=1\cdot41)$ can have only *three* integral or zero components along a reference direction (here assumed to be from top to bottom in the plane of the paper): $+1, 0$ and -1. Thus the angular momentum vector in this instance can be oriented in only three different directions (Fig. 2.5(b)–(d)) with respect to the reference direction. All three

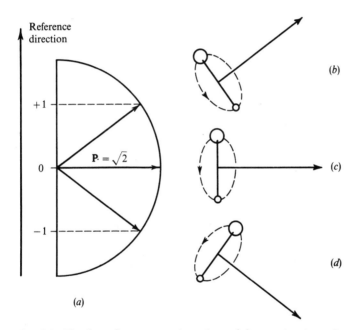

Fig. 2.5: *The three degenerate orientations of the rotational angular momentum vector for a molecule with $J=1$.*

rotational directions are, of course, associated with the same angular momentum and hence the same rotational energy: the $J=1$ level is thus three-fold degenerate.

Figure 2.6(a) and (b) shows the situation for $J=2$ $(\mathbf{P}=\sqrt{6})$

and $J=3$ ($\mathbf{P}=2\sqrt{3}$) with five-fold and seven-fold degeneracy respectively.   In general it may readily be seen that *each energy level
is $2J+1$-fold degenerate.*

Thus we see that, although the molecular population in each level
decreases exponentially (equation (2.19)), the number of degenerate
levels available increases rapidly with J.   The total relative population at an energy E_J will plainly be:

$$\text{population} \propto (2J+1) \exp(-E_J/kT). \tag{2.21}$$

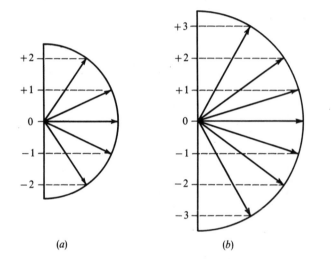

(a)   (b)

*Fig. 2.6: The five and seven degenerate rotational orientations for a molecule
with $J=2$ and $J=3$ respectively.*

When this is plotted against J the points fall on a curve of the type
shown in Fig. 2.7, indicating that the population rises to a maximum
and then diminishes.   Differentiation of equation (2.21) shows that
the population is a maximum at the nearest integral J value to:

$$\text{maximum population:} \quad J = \sqrt{\frac{kT}{2hcB}} - \frac{1}{2}. \tag{2.21a}$$

We have seen that line intensities are directly proportional to the
populations of the rotational levels, hence it is plain that transitions
between levels with very low or very high J values will have small
intensities while the intensity will be a maximum at or near the J
value given by equation (2.21a).

3.3. The Effect of Isotopic Substitution.

When a particular atom in a molecule is replaced by its isotope—an element identical in every way except for its atomic mass—the resulting substance is identical chemically with the original. In particular there is no appreciable change in internuclear distance on isotopic substitution.

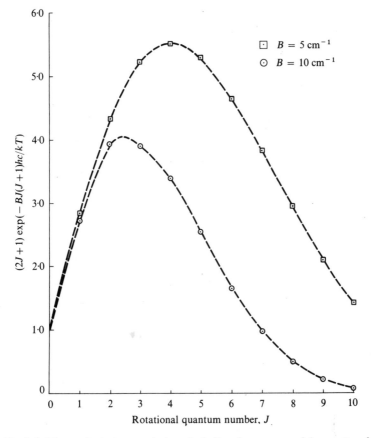

Fig. 2.7: The total relative populations, including degeneracy, of the rotational energy levels of a diatomic molecule; the diagram has been drawn for the same conditions as Fig. 2.4.

There is, however, a change in total mass and hence in the moment of inertia and B value for the molecule.

Considering carbon monoxide as an example, we see that on going from $^{12}C^{16}O$ to $^{13}C^{16}O$ there is a mass increase and hence a decrease in the B value. If we designate the ^{13}C molecule with a prime we have $B > B'$. This change will be reflected in the rotational energy levels of the molecule and Fig. 2.8 shows, much exaggerated, the relative lowering of the ^{13}C levels with respect to

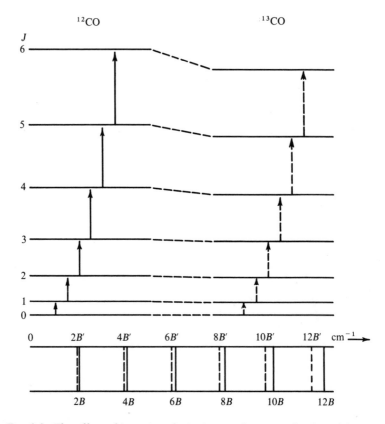

Fig. 2.8: The effect of isotopic substitution on the energy levels and hence rotational spectrum of a diatomic molecule such as carbon monoxide.

those of ^{12}C. Plainly, as shown by the diagram at the foot of this figure, the spectrum of the heavier species will show a smaller separation between the lines $(2B')$ than that of the lighter one $(2B)$. Again the effect has been much exaggerated for clarity, and the transitions due to the heavier molecule are shown dotted.

Observation of this decreased separation has led to the evaluation of precise atomic weights. Gilliam et al., as already stated, found the first rotational absorption of $^{12}C^{16}O$ to be at 3·84235 cm^{-1}, while that of $^{13}C^{16}O$ was at 3·67337 cm^{-1}. The values of B determined from these figures are:

$$B = 1·92118 \text{ cm}^{-1} \quad \text{and} \quad B' = 1·83669 \text{ cm}^{-1},$$

where the prime refers to the heavier molecule. We have immediately:

$$\frac{B}{B'} = \frac{h}{8\pi^2 Ic} \cdot \frac{8\pi^2 I'c}{h} = \frac{I'}{I} = \frac{\mu'}{\mu} = 1·046,$$

where μ is the reduced mass, and the internuclear distance is considered unchanged by isotopic substitution. Taking the mass of oxygen to be 15·9949 and that of carbon-12 to be 12·00, we have:

$$\frac{\mu'}{\mu} = 1·046 = \frac{15·9949m'}{15·9949 + m'} \times \frac{12 + 15·9949}{12 \times 15·9949}$$

from which m', the atomic weight of carbon-13, is found to be 13·0007. This is within 0·02% of the best value obtained in other ways.

It is noteworthy that the data quoted above were obtained by Gilliam *et al.* from $^{13}C^{16}O$ molecules in natural abundance (i.e. about 1% of ordinary carbon monoxide). Thus, besides allowing an extremely precise determination of atomic weights, microwave studies can give directly an estimate of the abundance of isotopes by comparison of absorption intensities.

3.4. The Non-Rigid Rotator. At the end of Section 3.1 we indicated how internuclear distances could be calculated from microwave spectra. It must be admitted that we selected our data carefully at this point—spectral lines for carbon monoxide, other than the first, would not have shown the constant $2B$ separation predicted by equation (2.16). This is shown by the spectrum of hydrogen fluoride given in Table 2.1; it is evident that the separation between successive lines (and hence the apparent B value) decreases steadily with increasing J.

The reason for this decrease may be seen if we calculate internuclear distances from the B values. The calculations are exactly similar to those of Section 3.1 and the results are shown in column 5 of Table 2.1. Plainly the bond length increases with J and we can see that our assumption of a *rigid* bond is only an approximation;

4

in fact, of course, all bonds are elastic to some extent, and the increase in length with J merely reflects the fact that the more quickly a diatomic molecule rotates the greater is the centrifugal force tending to move the atoms apart.

TABLE 2.1: *Microwave Spectrum of Hydrogen Fluoride*

J	$\bar{\nu}_{obs.}$ * (cm^{-1})	$\bar{\nu}_{calc.}$† (cm^{-1})	$\Delta\bar{\nu}_{obs.}$ (cm^{-1})	B ($=\frac{1}{2}\Delta\bar{\nu}$)	r (Å)
0	41·08	41·11			
			41·11	20·56	0·929
1	82·19	82·18			
			40·96	20·48	0·931
2	123·15	123·14			
			40·85	20·43	0·932
3	164·00	163·94			
			40·62	20·31	0·935
4	204·62	204·55			
			40·31	20·16	0·938
5	244·93	244·89			
			40·08	20·04	0·941
6	285·01	284·93			
			39·64	19·82	0·946
7	324·65	324·61			
			39·28	19·64	0·951
8	363·93	363·89			
			38·89	19·45	0·955
9	402·82	402·70			
			38·31	19·16	0·963
10	441·13	441·00			
			37·81	18·91	0·969
11	478·94	478·74			

* Lines numbered according to $\bar{\nu}_J = 2B(J+1)$ cm^{-1}. Observed data from "An Examination of the Far Infra-red Spectrum of Hydrogen Fluoride" by A. A. Mason and A. H. Nielsen, published as Scientific Report No. 5, August 1963, Contract No. AF 19(604)-7981, by kind permission of the authors.
† See Section 3.5 for details of the calculation.

Before showing how this elasticity may be quantitatively allowed for in rotational spectra, we shall consider briefly two of its consequences. Firstly, when the bond is elastic, a molecule may have vibrational energy—i.e. the bond will stretch and compress periodically with a certain fundamental frequency dependent upon the masses of the atoms and the elasticity (or force constant) of the bond. If the motion is simple harmonic (which, we shall see in Chapter 3, is usually a very good approximation to the truth) the force constant is given by:

$$k = 4\pi^2 \bar{\omega}^2 c^2 \mu, \tag{2.22}$$

where $\bar{\omega}$ is the vibration frequency (expressed in cm^{-1}), and c and μ have their previous definitions. Plainly the variation of B with J is determined by the force constant—the weaker the bond, the more readily will it distort under centrifugal forces.

The second consequence of elasticity is that the quantities r and B vary during a vibration. When these quantities are measured by microwave techniques many hundreds of vibrations occur during a rotation, and hence the measured value is an average. However, from the defining equation of B we have:

$$B = \frac{h}{8\pi^2 I c} = \frac{h}{8\pi^2 c \mu r^2}$$

or

$$B \propto 1/r^2 \tag{2.23}$$

since all other quantities are independent of vibration. Now, although in simple harmonic motion a molecular bond is compressed and extended an equal amount on each side of the equilibrium distance (see Fig. 2.9) and the average value of the distance is therefore unchanged, the average value of $1/r^2$ is *not* equal to $1/r_0^2$, where r_0 is the equilibrium distance. We can see this most easily by an example. Consider a bond of equilibrium length 1 Å vibrating between the limits 0·9 and 1·1 Å. We have:

$$\langle r \rangle_{\text{av.}} = \frac{0·9 + 1·1}{2} = 1·0 = r_0$$

but

$$\left\langle \frac{1}{r^2} \right\rangle_{\text{av.}} = \frac{(1/0·9)^2 + (1/1·1)^2}{2} = 1·03 \text{ Å.}$$

The difference, though small, is not negligible compared with the precision with which B can be measured spectroscopically.

It is usual, then, to define three different sets of values for B and r. At the equilibrium separation, r_e, between the nuclei, the rotational constant is B_e; in the vibrational ground state the average internuclear separation is r_0 associated with a rotational constant B_0;

while if the molecule has excess vibrational energy the quantities are r_n and B_n, where n is the vibrational quantum number.

During the remainder of this chapter we shall ignore the small differences between B_0, B_e and B_n—the discrepancy is most important in the consideration of vibrational spectra in Chapter 3.

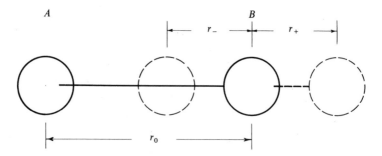

Fig. 2.9: *Showing the equal compression and extension of a bond undergoing simple harmonic motion.*

3.5. *The Spectrum of a Non-Rigid Rotator.* The Schrödinger wave equation may be set up for a non-rigid molecule, and the rotational energy levels are found to be:

$$E_J = \frac{h^2}{8\pi^2 I} J(J+1) - \frac{h^4}{32\pi^4 I^2 r^2 k} J^2(J+1)^2 \text{ ergs}$$

or

$$\varepsilon_J = E_J/hc = BJ(J+1) - DJ^2(J+1)^2 \text{ cm}^{-1}, \qquad (2.24)$$

where the rotational constant, B, is as defined previously, and the *centrifugal distortion constant* D, is given by:

$$D = \frac{h^3}{32\pi^4 I^2 r^2 kc} \text{ cm}^{-1} \qquad (2.25)$$

which is a positive quantity. Equation (2.24) applies for a simple harmonic force field only; if the force field is anharmonic, the expression becomes:

$$\varepsilon_J = BJ(J+1) - DJ^2(J+1)^2 + HJ^3(J+1)^3 + KJ^4(J+1)^4 \dots \text{ cm}^{-1},$$

$$(2.26)$$

where H, K, etc., are small constants dependent upon the geometry of the molecule. They are, however, negligible compared with D and most modern spectroscopic data is adequately fitted by equation (2.24).

From the defining equations of B and D it may be shown directly that

$$D = \frac{16B^3\pi^2\mu c^2}{k} = \frac{4B^3}{\bar{\omega}^2} \qquad (2.27)$$

where $\bar{\omega}$ is the vibrational frequency of the bond, and k has been expressed according to equation (2.22). We shall see in Chapter 3 that vibrational frequencies are usually of the order of 10^3 cm^{-1}, while B we have found to be of the order of 10 cm^{-1}. Thus we see that D, being of the order 10^{-3} cm^{-1}, is very small compared with B. For small J, therefore, the correction term $DJ^2(J+1)^2$ is almost negligible, while for J values of 10 or more it may become appreciable.

Figure 2.10 shows, much exaggerated, the lowering of rotational levels when passing from the rigid to the non-rigid diatomic molecule. The spectra are also compared, the dotted lines connecting corresponding energy levels and transitions of the rigid and the non-rigid molecules. It should be noted that the selection rule for the latter is still $\Delta J = \pm 1$.

We may easily write an analytical expression for the transitions:

$$\begin{aligned}
\varepsilon_{J+1} - \varepsilon_J = \bar{v}_J &= B[(J+1)(J+2) - J(J+1)] \\
&\quad - D[(J+1)^2(J+2)^2 - J^2(J+1)^2] \\
&= 2B(J+1) - 4D(J+1)^3 \text{ cm}^{-1}, \qquad (2.28)
\end{aligned}$$

where $\bar{v}_J$ represents equally the upward transition from J to $J+1$, or the downward from $J+1$ to J. Thus we see analytically, and from Fig. 2.10, that the spectrum of the elastic rotor is similar to that of the rigid molecule except that each line is displaced slightly to low frequency, the displacement increasing with $(J+1)^3$.

A knowledge of D gives rise to two useful items of information. Firstly, it allows us to determine the J value of lines in an observed spectrum. If we have measured a few transitions there is no *a priori* way of determining from which J value they arise; however, fitting equation (2.28) to them—provided three consecutive lines have been measured—gives unique values for B, D and J. The precision of such fitting is shown by Table 2.1 where the wave-numbers are calculated from the equation:

$$\bar{v}_J = 41{\cdot}122(J+1) - 8{\cdot}52 \times 10^{-3}(J+1)^3 \text{ cm}^{-1}. \qquad (2.29)$$

Secondly, a knowledge of D enables us to determine—although

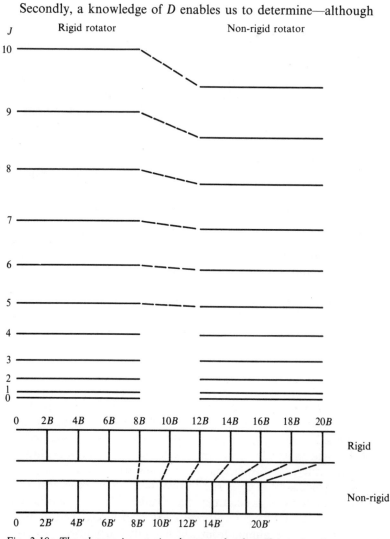

Fig. 2.10: The change in rotational energy levels and rotational spectrum when passing from a rigid to a non-rigid diatomic molecule. Levels on the right calculated using $D = 10^{-3}B$.

rather inaccurately—the vibrational frequency of a diatomic
molecule. From the above data for hydrogen fluoride and equation
(2.27) we have:

$$\bar{\omega}^2 = \frac{4B^3}{D} = 16.33 \times 10^6 \ (\text{cm}^{-1})^2,$$

i.e.

$$\bar{\omega} \approx 4050 \ \text{cm}^{-1}.$$

In the next chapter we shall see that a more precise determination
leads to the value 4138.3 cm^{-1}; the 2% inaccuracy in the present
calculation is due partly to the assumption of simple harmonic
motion, and partly to the very small, and hence relatively inaccurate,
value of D.

The force constant follows directly:

$$k = 4\pi^2 c^2 \bar{\omega}^2 \mu = 9.6 \times 10^5 \ \text{dynes/cm},$$

which indicates, as expected, that H—F is a relatively strong bond.

4. Polyatomic Molecules

4.1. Linear Molecules. We consider first molecules such as
carbon oxysulphide OCS, or chloroacetylene HC≡CCl, where all
the atoms lie on a straight line, since this type gives rise to particularly simple spectra in the microwave region. Since $I_B = I_C$;
$I_A = 0$, as for diatomic molecules, the energy levels are given by a
formula identical with equation (2.26), i.e.:

$$\varepsilon_J = BJ(J+1) - DJ^2(J+1)^2 + \cdots \ \text{cm}^{-1}, \tag{2.30}$$

and the spectrum will show the same $2B$ separation modified by the
distortion constant. In fact, the whole of the discussion on diatomic molecules applies equally to all linear molecules; three
points, however, should be underlined:

(*i*) Since the moment of inertia for the end-over-end rotation of a
polyatomic linear molecule is considerably greater than that of a
diatomic molecule, the B value will be much smaller, and the
spectral lines more closely spaced. Thus B values for diatomic
molecules are about 10 cm^{-1}, while for triatomic molecules they
can be 1 cm^{-1} or less, and for larger molecules smaller still.

(*ii*) The molecule must, as usual, possess a dipole moment if it is
to exhibit a rotational spectrum. Thus OCS will be microwave

active, while OCO (more usually written CO_2) will not. In particular, it should be noted that isotopic substitution does not lead to a dipole moment since the bond lengths and atomic charges are unaltered by the substitution. Thus $^{16}OC^{18}O$ is microwave inactive.

(*iii*) A non-cyclic polyatomic molecule containing N atoms has altogether $N-1$ individual bond lengths to be determined. Thus in the triatomic molecule OCS there is the CO distance, r_{CO}, and the CS distance, r_{CS}. On the other hand, there is only *one* moment of inertia for the end-over-end rotation of OCS, and only this one value can be determined from the spectrum. Table 2.2 shows the data for this molecule. Over the four lines observed there is seen

TABLE 2.2: Microwave Spectrum of Carbon Oxysulphide

J	$\bar{\nu}_{obs.}$ (cm^{-1})	$\Delta\bar{\nu}$	B (cm^{-1})
0	—		
		2×0.4055	0.2027
1	0.8109		
		0.4054	0.2027
2	1.2163		
		0.4054	0.2027
3	1.6217		
		0.4054	0.2027
4	2.0271		

to be no appreciable centrifugal distortion, and, taking the value of B as 0.2027 cm^{-1}, we calculate:

$$I_B = \frac{h}{8\pi^2 Bc} = 137.95 \times 10^{-40} \text{ g cm}^2.$$

From this one observation it is plainly impossible to deduce the two unknowns, r_{CO} and r_{CS}. The difficulty can be overcome, however, if we study a molecule with different atomic masses but the *same* bond lengths—i.e. an isotopically substituted molecule—since this will have a different moment of inertia.

Let us consider the rotation of OCS in some detail. Figure 2.11 shows the molecule, where r_O, r_C and r_S represent the distances of the atoms from the centre of gravity. Consideration of moments gives:

$$m_O r_O + m_C r_C = m_S r_S, \qquad (2.31)$$

where m_i is the mass of atom i. The moment of inertia is:

$$I = m_\mathrm{O} r_\mathrm{O}^2 + m_\mathrm{C} r_\mathrm{C}^2 + m_\mathrm{S} r_\mathrm{S}^2 \tag{2.32}$$

and we have the further equations:

$$r_\mathrm{O} = r_\mathrm{CO} + r_\mathrm{C}; \qquad r_\mathrm{S} = r_\mathrm{CS} - r_\mathrm{C}, \tag{2.33}$$

where r_CO and r_CS are the bond lengths of the molecule. It is these

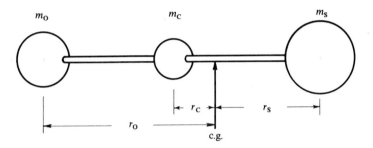

Fig. 2.11: *The molecule of carbon oxysulphide*, OCS, *showing the distances of each atom from the centre of gravity.*

we wish to determine. Substituting (2.33) in (2.31) and collecting terms:

$$(m_\mathrm{C} + m_\mathrm{O} + m_\mathrm{S}) r_\mathrm{C} = m_\mathrm{S} r_\mathrm{CS} - m_\mathrm{O} r_\mathrm{CO}$$

or

$$M r_\mathrm{C} = m_\mathrm{S} r_\mathrm{CS} - m_\mathrm{O} r_\mathrm{CO}, \tag{2.34}$$

where we write M for the total mass of the molecule. Substituting (2.33) in (2.32):

$$I = m_\mathrm{O} (r_\mathrm{CO} + r_\mathrm{C})^2 + m_\mathrm{C} r_\mathrm{C}^2 + m_\mathrm{S} (r_\mathrm{CS} - r_\mathrm{C})^2$$
$$= M r_\mathrm{C}^2 + 2 r_\mathrm{C} (m_\mathrm{O} r_\mathrm{CO} - m_\mathrm{S} r_\mathrm{CS}) + m_\mathrm{O} r_\mathrm{CO}^2 + m_\mathrm{S} r_\mathrm{CS}^2$$

and finally substituting for r_C from equation (2.34):

$$I = m_\mathrm{O} r_\mathrm{CO}^2 + m_\mathrm{S} r_\mathrm{CS}^2 - \frac{(m_\mathrm{O} r_\mathrm{CO} - m_\mathrm{S} r_\mathrm{CS})^2}{M}. \tag{2.35}$$

Considering now the isotopic molecule, ^{18}OCS, we may write m_O' for m_O throughout equation (2.35):

$$I' = m_\mathrm{O}' r_\mathrm{CO}^2 + m_\mathrm{S} r_\mathrm{CS}^2 - \frac{(m_\mathrm{O}' r_\mathrm{CO} - m_\mathrm{S} r_\mathrm{CS})^2}{M'} \tag{2.36}$$

and we can now solve for r_{CO} and r_{CS}, provided we have extracted a value for I' from the microwave spectrum of the isotopic molecule. Note that we do *not* need to write r'_{CO}, since we assume that the bond length is unaltered by isotopic substitution. This assumption may be checked by studying the molecules $^{16}OC^{34}S$ and $^{18}OC^{34}S$, since we would then have four moments of inertia. The bond distances found are quite consistent, and hence justify the assumption.

The extension of the above discussion to molecules with more than three atoms is straightforward; it suffices to say here that microwave studies have led to very precise determinations of many bond lengths in such molecules.

4.2. Symmetric Top Molecules. Although the rotational energy levels of this type of molecule are more complicated than those of linear molecules, we shall see that, because of their symmetry, their pure rotational spectra are still relatively simple. Choosing methyl fluoride again as our example we remember that

$$I_B = I_C \neq I_A; \qquad I_A \neq 0.$$

There are now two directions of rotation in which the molecule can absorb or emit energy—that about the main symmetry axis (the C—F bond in this case) and that perpendicular to this axis.

We thus need two quantum numbers to describe the degree of rotation, one for I_A and one for I_B or I_C. However, it turns out to be very convenient mathematically to have a quantum number to represent the *total* angular momentum of the molecule, which is the sum of the separate angular momenta about the two different axes. This is usually chosen to be the quantum number J. Reverting for a moment to *linear* molecules, remember that we there used J to represent the end-over-end rotation of a molecule: however, this was the *only* sort of rotation allowed, so it is quite consistent to use J, in general, to represent the *total angular momentum*. It is then conventional to use K to represent the angular momentum about the top axis—i.e. about the C—F bond in this case.

Let us briefly consider what values are allowed to K and J. Both must, by the conditions of quantum mechanics, be integral or zero. The total angular momentum can be as large as we like, i.e. J can be 0, 1, 2,..., ∞ (except, of course, for the theoretical possibility that a real molecule will be disrupted at very high rotational speeds). Once we have chosen J, however, K is rather more limited. Let us consider the case when $J = 3$. Plainly the rotational energy can be divided in several ways between motion about the main symmetry axis and motion perpendicular to this. If *all* the rotation is about the axis, $K = 3$; but note that K cannot be greater than J since J is the *total* angular momentum. Equally we could have $K = 2$, 1 or 0, in which case the motion perpendicular to the axis would have 1, 2 or 3 respectively. Additionally, however, K can be negative—we can imagine positive and negative values of K to correspond with clockwise and anticlockwise rotation about the symmetry axis—and so can have values -1, -2 or -3.

In general, then, for a total angular momentum J, we see that K can take values:

$$K = J, J-1, J-2, \ldots, 0, \ldots, -(J-1), -J \qquad (2.37)$$

which is a total of $2J + 1$ values altogether. This figure of $2J + 1$ is important and will recur.

If we take first the case of a *rigid* symmetric top—i.e. one in which the bonds are supposed not to stretch under centrifugal forces—the Schrödinger equation may be solved to give the allowed energy levels for rotation as:

$$\varepsilon_{J,K} = E_{J,K}/hc = BJ(J+1) + (A-B)K^2 \text{ cm}^{-1} \qquad (2.38)$$

where, as before,

$$B = \frac{h}{8\pi^2 I_B c} \quad \text{and} \quad A = \frac{h}{8\pi^2 I_A c}.$$

Note that the energy depends on K^2, so that it is immaterial whether the top spins clockwise or anticlockwise: the energy is the same for a given angular momentum. For all $K > 0$, therefore, the rotational energy levels are *doubly degenerate*.

The selection rules for this molecule may be shown to be:

$$\Delta J = \pm 1 \text{ (as before)}, \quad \text{and} \quad \Delta K = 0 \qquad (2.39)$$

and, when these are applied to equation (2.38), the spectrum is given by:

$$\begin{aligned}
\varepsilon_{J+1,K} - \varepsilon_{J,K} = \bar{\nu}_{J,K} &= B(J+1)(J+2) + (A-B)K^2 \\
&\quad - [BJ(J+1) + (A-B)K^2] \\
&= 2B(J+1) \text{ cm}^{-1}.
\end{aligned} \qquad (2.40)$$

Thus the spectrum is independent of K, and hence rotational changes about the symmetry axis do not give rise to a rotational spectrum. The reason for this is quite evident—rotation about the symmetry axis does not change the dipole moment perpendicular to the axis (which always remains zero), and hence the rotation cannot interact with radiation. Equation (2.40) shows that the spectrum is just the same as for a linear molecule and that only one moment of inertia—that for end-over-end rotation—can be measured.

Both equations (2.38) and (2.40) are for a rigid molecule, however, and we have already seen that microwave spectroscopy is well able to detect the departure of real molecules from this idealized state. When centrifugal stretching is taken into account, the energy levels become:

$$\varepsilon_{J,K} = BJ(J+1)+(A-B)K^2-D_J J^2(J+1)^2$$
$$-D_{JK}J(J+1)K^2-D_K K^4 \text{ cm}^{-1} \quad (2.41)$$

where, in an obvious notation, D_J, D_{JK} and D_K are small correction terms for non-rigidity. The selection rules are unchanged (equation (2.39)), and so the spectrum is:

$$\bar{v}_{J,K} = \varepsilon_{J+1,K}-\varepsilon_{J,K}$$
$$= 2B(J+1)-4D_J(J+1)^3-2D_{JK}(J+1)K^2 \text{ cm}^{-1}. \quad (2.42)$$

We see that the spectrum will be basically that of a linear molecule (including centrifugal stretching), with an additional term depending on K^2.

It is easy to see why this spectrum now depends on the axial rotation (i.e. depends on K), although such rotation produces no dipole change. Figure 2.12 illustrates methyl fluoride for (a) $K=0$, no axial rotation, and (b) $K=1$, the molecule rotating about the symmetry axis. We see, from the much exaggerated diagram, that axial rotation widens the HCH angles and stretches the C—H bonds. The distorted molecule (b) has a different moment of inertia for end-over-end rotation than (a). If we write equation (2.42) as:

$$\bar{v}_{JK} = 2(J+1)[B-2D_J(J+1)^2-D_{JK}K^2] \text{ cm}^{-1}$$

we can see more clearly that the centrifugal distortion constants D_J

and D_{JK} can be considered as correction terms to the rotational constant B, and hence as perturbing the moment of inertia I_B.

Since each value of J is associated with $2J+1$ values of K, we see that each line characterized by a certain J value must have $2J+1$

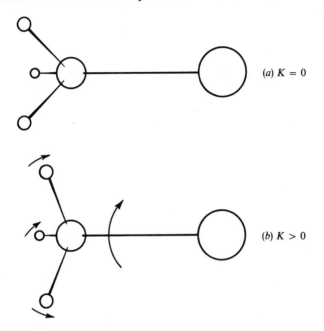

Fig. 2.12: *The influence of axial rotation on the moment of inertia of a symmetric top molecule, e.g. methyl fluoride,* CH_3F. *In* (a) *there is no axial rotation* ($K=0$), *and in* (b) $K>0$.

components. However, since K only appears as K^2 in equation (2.42), there will be only $J+1$ *different* frequencies, all those with $K>0$ being doubly degenerate. We may tabulate a few lines as follows:

$$J = 0, K = 0 \qquad \bar{\nu}_{JK} = 2B-4D_J \text{ cm}^{-1}$$

$$J = 1, K = 0 \qquad \bar{\nu}_{JK} = 4B-32D_J$$

$$K = \pm 1 \qquad \bar{\nu}_{JK} = 4B-32D_J-4D_{JK} \qquad (2.43)$$

$$J = 2, K = 0 \qquad \bar{\nu}_{JK} = 6B-108D_J$$

$$K = \pm 1 \qquad \bar{\nu}_{JK} = 6B-108D_J-6D_{JK}$$

$$K = \pm 2 \qquad \bar{\nu}_{JK} = 6B-108D_J-24D_{JK}, \text{ etc.}$$

Let us now compare these with the observed spectrum of methyl fluoride. This is shown as a line diagram in Fig. 2.13, and the

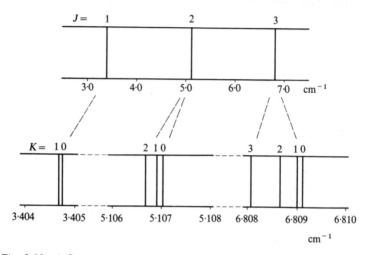

Fig. 2.13: A diagrammatic representation of the rotational spectrum of the symmetric top molecule methyl fluoride, CH_3F.

frequencies are tabulated in Table 2.3. Fitting these data to equations such as (2.43) leads directly to:

$$B = 0.851204 \text{ cm}^{-1}$$

$$D_J = 2.00 \times 10^{-6} \text{ cm}^{-1}$$

$$D_{JK} = 1.47 \times 10^{-5} \text{ cm}^{-1}.$$

The calculated frequencies of Table 2.3 show how precisely such measurements may now be made.

Once again each spectrum examined yields only one value of B, but the spectra of isotopic molecules can, in principle, give sufficient information for the calculation of all the bond lengths and angles of symmetric top molecules, together with estimates of the force constant of each bond.

4.3. Asymmetric Top Molecules. Since spherical tops show no microwave spectrum (cf. Section 1(*iii*)), the only other class of molecule of interest here is the asymmetric top which has (Section 1(*iv*)) all three moments of inertia different. These molecules will not detain us long since their rotational energy levels and spectra are very complex—in fact, no analytical expressions can be written for them corresponding to equations (2.24) and (2.28) for linear or equations (2.41) and (2.42) for symmetric top molecules. Each molecule and spectrum must, therefore, be treated as a separate case, and much tedious computation is necessary before structural

TABLE 2.3: Microwave Spectrum of Methyl Fluoride

J	K	$\bar{v}_{obs.}$*(cm^{-1})	$\bar{v}_{calc.}$ (cm^{-1})
1	0	3·40475	3·404752
	1	3·40470	3·404693
2	0	5·10701	5·107008
	1	5·10692	5·106920
	2	5·10665	5·106655
3	0	6·80912	6·809120
	1	6·80900	6·809002
	2	6·80865	6·808649
	3	6·80806	6·808062

* Taken from W. Gordy, *Physical Review*, **93**, 406 (1954), by kind permission of the author.

parameters can be determined. The best method of attack so far has been to consider the asymmetric top as falling somewhere between the oblate and prolate symmetric top; interpolation between the two sets of energy levels of the latter leads to a first approximation of the energy levels—and hence spectrum—of the asymmetric molecule. It suffices to say that arbitrary methods such as this have been quite successful, and much very precise structural data has been published.

In order to give an idea of the precision of such measurements, we collect in Table 2.4 some molecular data determined by microwave methods, including examples from diatomic and linear molecules, symmetric tops, and asymmetric tops.

5. Techniques and Instrumentation

5.1. It is not proposed here to give more than a brief outline of the techniques of microwave spectroscopy since detailed accounts are available in some of the books listed in the bibliography.

TABLE 2.4: Some Molecular Data Determined by Microwave Spectroscopy

Molecule	Type	Bond Length (Å)	Bond Angle	Dipole Moment* (10^{-18} e.s.u.)
NaCl	Diatomic	$2\cdot3606 \pm 0\cdot0001$	—	$8\cdot5 \pm 0\cdot2$
COS	Linear	$\begin{cases} 1\cdot164 \pm 0\cdot001 \text{ (C—O)} \\ 1\cdot559 \pm 0\cdot001 \text{ (C—S)} \end{cases}$	—	$0\cdot712 \pm 0\cdot004$
HCN	Linear	$\begin{cases} 1\cdot06317 \pm 0\cdot00005 \text{ (C—H)} \\ 1\cdot15535 \pm 0\cdot00006 \text{ (C—N)} \end{cases}$	—	$2\cdot986 \pm 0\cdot004$
NH_3	Sym. Top	$1\cdot008 \pm 0\cdot004$	$107\cdot3° \pm 0\cdot2°$	$1\cdot47 \pm 0\cdot01$
CH_3Cl	Sym. Top	$\begin{cases} 1\cdot0959 \pm 0\cdot0005 \text{ (C—H)} \\ 1\cdot7812 \pm 0\cdot0005 \text{ (C—Cl)} \end{cases}$	$108\cdot0° \pm 0\cdot2°$ (HCH)	$1\cdot871 \pm 0\cdot005$
H_2O	Asym. Top	$0\cdot9584 \pm 0\cdot0005$	$104\cdot5° \pm 0\cdot3°$	$1\cdot846 \pm 0\cdot005$
O_3	Asym. Top	$1\cdot278 \pm 0\cdot002$	$116\cdot8° \pm 0\cdot5°$	$0\cdot53 \pm 0\cdot02$

* Measured from the Stark effect, cf. Section 5.2.

Microwave spectroscopy, of course, follows the usual pattern: source, monochromator, beam direction, sample and detector. We shall discuss each in turn.

(i) *The Source and Monochromator.* The usual source in this region is the klystron valve which, since it emits radiation of only a very narrow frequency range, is called "monochromatic" and acts as its own monochromator. The actual emission frequency is variable electronically and hence a spectrum may be scanned over a limited range of frequencies using a single klystron.

One slight disadvantage of this source is that the total energy radiated is very small—of the order of milliwatts only. However, since all this is concentrated into a narrow frequency band a sharply tuned detector can be sufficiently activated to produce a strong signal.

(ii) *Beam Direction.* This is achieved by the use of "wave-guides"—hollow tubes of copper or silver, usually of rectangular cross-section—inside which the radiation is confined. The wave-guides may be gently tapered or bent to allow focusing and directing of the radiation. Atmospheric absorption of the beam is consider-able, so the system must be efficiently evacuated.

(iii) *Sample and Sample Space.* In almost all microwave studies so far the sample has been gaseous. However, pressures of 0·01 mm Hg are sufficient to give a reasonable absorption spectrum, so many substances which are usually thought of as solid or liquid may be examined provided their vapour pressures are above this value. The sample is retained by very thin mica windows in a piece of evacuated wave-guide.

(*iv*) *Detector.* It is possible to use an ordinary superheterodyne radio receiver as detector, provided this may be tuned to the appropriate high frequency; however, a simple crystal detector is found to be more sensitive and easier to use. This detects the radiation focused upon it by the wave-guide, and the signal it gives is amplified electronically for display on an oscilloscope, or for permanent record on paper.

5.2. The Stark Effect. We cannot leave the subject of microwave spectroscopy without a brief description of the Stark effect and its applications. A more detailed discussion is to be found in the book by Townes and Schawlow mentioned in the bibliography.

Experimentally the Stark effect requires the placing of an electric field, either perpendicular or parallel to the direction of the radiation beam, across the sample. Practically it is simpler to have a perpendicular field. We shall consider three advantages of this field.

(*i*) A molecule exhibiting a rotational spectrum must have an electric dipole moment, and so its rotational energy levels will be perturbed by the application of an exterior field since interaction will occur. Put simply, the absorption lines of the spectrum will be shifted by an amount depending on the extent of the interaction, and thus depending on both E, the applied field, and μ the dipole moment. For a linear molecule the shift is found to be:

$$\Delta v \propto (\mu E)^2 \quad \text{(linear molecule)}$$

while for a symmetric top:

$$\Delta v \propto \mu E \quad \text{(symmetric top)}.$$

Thus we have immediately a very accurate method of determining dipole moments, simply by observation of the Stark shift. More important, the measurement is made on very dilute gas samples, so the dipole moment observed may be taken to be that of the actual molecule, uncomplicated by molecular interactions, solvent effects, etc. Some values determined in this way are included in Table 2.4.

(*ii*) The second valuable application of the Stark effect is in the assignment of observed spectral lines to particular J values. We have seen that, in the absence of marked departure from rigidity and good resolving power, the assignment of J values is not always

5

obvious. The line of lowest frequency which we observe *may* happen to correspond with $J = 0$, or it may be that it is the first observable line of a series, either because earlier lines are intrinsically very weak or because of limitations in the apparatus used. However, we have seen that each line is $2J + 1$ degenerate because rotations can occur in $2J + 1$ orientations in space without violating quantum laws.   In the absence of any orienting effect these transitions have precisely the same frequency, but a Stark field constitutes an orienting effect, and splits the degeneracy; thus multiplet structure is observed for all lines with $J > 0$.   The number of components depends on J, and hence unambiguous assignments can be made.

(*iii*)   The final application is purely an instrumental one, but is especially interesting in that it has its counterpart in other spectral regions.   We have already referred to the concept of signal-to-noise ratio in Chapter 1; that part of the noise which arises from random fluctuations in the background radiation may be removed by modulating the beam by means of the Stark effect as explained below.

Imagine the application of a Stark field in a periodic manner such as the "square-wave" variation of Fig. 2.14; while the field is

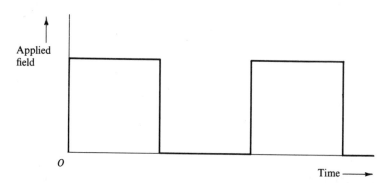

Fig. 2.14: A "square wave" potential as used for Stark modulation.

switched on the signal is modified in the way described in (*i*) and (*ii*) above.   If we arrange the modulation frequency to be some 100–1000 c/sec, and construct the amplifier so that it amplifies only the component of the signal which has the modulated frequency, stray radiation which has not been through the modulating field will be completely ignored. This results in a great improvement of the signal-to-noise ratio.

A further refinement is to arrange for both the modulated and unmodulated parts to be amplified separately and displayed on the same oscilloscope—the modulated part on the upper half, say, and the unmodulated on the lower. This much facilitates the measurement of the Stark splittings discussed in (*i*) and (*ii*) above.

6. Summary

We have seen how the two types of molecule discussed in detail—the linear and the symmetric top—give rise to the same basic pattern of spectral lines: a succession of absorption maxima separated by a nearly constant difference. Their spectra may be distinguished under high resolution since the symmetric top lines are multiplets with $J+1$ components, while lines from linear molecules have no fine structure. Asymmetric top molecules, on the other hand, give rise to more complex spectra which cannot be analysed in simple terms. Thus we have immediately a method for determining the overall shape of a given molecule.

If this were all that microwave spectroscopy could achieve, its usefulness would have been short-lived—it is not often that one does not know if a particular molecule is linear or not. However, we have seen that four further parameters are very accurately measureable from microwave spectra.

Firstly, the precise B values available can lead, after judicious isotopic substitution, to the bond lengths and angles of fairly simple molecules. Many accurate structural determinations have been made in this way.

Secondly, the departure of a molecule from the ideal rigid condition can be assessed by the evaluation of distortion constants such as D_J and D_{JK}. These lead to calculations of approximate molecular force constants.

Thirdly, measurements on isotopically substituted molecules have led to some of the most accurate estimates of relative atomic masses; and finally the Stark effect allows us to determine precise values of molecular dipole moments.

This all adds up to a large volume of useful structural information of great value to chemist and physicist alike.

Bibliography

BARROW, G. M.: *Introduction to Molecular Spectroscopy*, McGraw-Hill Book Co. Inc., 1962

BERL, W. G. (Ed.): *Physical Methods in Chemical Analysis*, Vol. III, Academic Press, 1956

BRAUDE, E. A. and F. C. NACHOD: *Determination of Organic Structures by Physical Methods*, Vol. 1, Academic Press, 1956

GORDY, W., W. V. SMITH and R. TRAMBARULO: *Microwave Spectroscopy*, John Wiley, 1953

KING, G. W.: *Spectroscopy and Molecular Structure*, Holt, Rinehart and Winston Inc., 1964

SQUIRES, T. L.: *Introduction to Microwave Spectroscopy*, George Newnes Ltd., 1964

STRANDBERG, M. W. P.: *Microwave Spectroscopy*, Methuen, 1954

TOWNES, C. H. and A. L. SCHAWLOW: *Microwave Spectroscopy*, McGraw-Hill Book Co. Inc., 1955

WALKER, S. and H. STRAW: *Spectroscopy*, Vol. I, Chapman and Hall, 1961

CHAPTER 3

INFRA-RED SPECTROSCOPY

We saw in the previous chapter how the elasticity of chemical bonds led to anomalous results in the rotational spectra of rapidly rotating molecules—the bonds stretched under centrifugal forces. In this chapter we consider another consequence of this elasticity— the fact that atoms in a molecule do not remain in fixed relative positions but vibrate about some mean position. We consider first the case of a diatomic molecule and the spectrum which arises if its only motion is vibration; then we shall deal with the more practical case of a diatomic molecule undergoing vibration and rotation simultaneously; finally we shall extend the discussion to more complex molecules.

1. The Vibrating Diatomic Molecule

1.1. The Energy of a Diatomic Molecule. When two atoms combine to form a stable covalent molecule (e.g. HCl gas) they may be said to do so because of some internal electronic rearrangement. This is not the place to discuss the detailed mechanisms of chemical bond formation; we may simply look on the phenomenon as a balancing of forces. On the one hand there is a repulsion between the positively charged nuclei of both atoms, and between the negative electron "clouds"; on the other there is an attraction between the nucleus of one atom and the electrons of the other, and vice versa. The two atoms settle at a mean internuclear distance such that these forces are just balanced and the energy of the whole system is at a minimum. Attempt to squeeze the atoms more closely together and the repulsive force rises rapidly; attempt to pull them further apart and we are resisted by the attractive force. In either case an attempt to distort the bond requires an input of energy and so we may plot energy against internuclear distance as in Fig. 3.1. At the minimum the internuclear distance is referred to as the equilibrium distance r_{eq}, or more simply, as the bond length.

The compression and extension of a bond may be likened to the

59

behaviour of a spring and we may extend the analogy by assuming that the bond, like a spring, obeys Hooke's law. We may then write

$$f = -k(r - r_{eq}),\qquad(3.1)$$

where f is the restoring force, k the force constant and r the internuclear distance. In this case the energy curve is parabolic and has the form

$$E = \tfrac{1}{2}k(r - r_{eq})^2.\qquad(3.2)$$

This model of a vibrating diatomic molecule—the so-called simple harmonic oscillator model—while only an approximation,

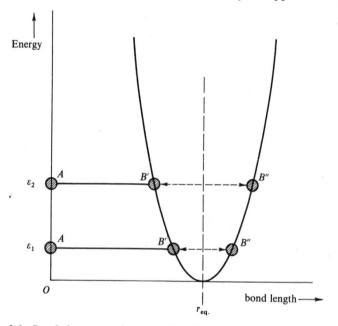

Fig. 3.1: *Parabolic curve of energy plotted against the extension or compression of a spring obeying Hooke's law.*

forms an excellent starting point for the discussion of vibrational spectra.

1.2. The Simple Harmonic Oscillator. In Fig. 3.1 we have plotted the energy according to equation (3.2). The zero of curve and equation is found at $r = r_{eq}$, and any energy in excess of this, e.g. ε_1, arises because of extension or compression of the bond. The figure shows that if one atom (A) is considered to be stationary on the $r = 0$ axis, the other will oscillate between B' and B''. If the energy is increased to ε_2 the oscillation will become more vigorous —that is to say the degree of compression or extension will be greater—but the vibrational frequency will not change. An elastic bond, like a spring, has a certain vibration frequency dependent upon the mass of the system and the force constant, but independent of the amount of distortion. Classically it is easy to show that the oscillation frequency is:

$$\omega_{\text{osc.}} = \frac{1}{2\pi} \sqrt{\frac{k}{\mu}} \text{ c/sec,} \tag{3.3}$$

where μ is the reduced mass of the system (cf. equation (2.9)). If the force constant k is expressed in dynes/cm, equation (3.3) gives the classical frequency, ω_{osc}, in c/sec. To convert to cm^{-1} (cf. Chapter 1) we must write the classical oscillation frequency as:

$$\bar{\omega}_{\text{osc.}} = \frac{1}{2\pi c} \sqrt{\frac{k}{\mu}} \text{ cm}^{-1}. \tag{3.4}$$

Vibrational energies, like all other molecular energies, are quantized, and the allowed vibrational energies for any particular system may be calculated from the Schrödinger equation. For the simple harmonic oscillator these turn out to be:

$$E_v = (v + \tfrac{1}{2})h\omega_{\text{osc.}} \text{ ergs} \quad (v = 0, 1, 2, \ldots), \tag{3.5}$$

where v is called the *vibrational quantum number*. Converting to the spectroscopic units, cm^{-1}, we have:

$$\varepsilon_v = \frac{E_v}{hc} = (v + \tfrac{1}{2})\bar{\omega}_{\text{osc.}} \text{ cm}^{-1} \tag{3.6}$$

as the only energies allowed to a simple harmonic vibrator. Some of these are shown in Fig. 3.2.

In particular we should notice that the *lowest* vibrational energy, obtained by putting $v = 0$ in equation (3.5) or (3.6), is

$$E_0 = \tfrac{1}{2}h\omega_{\text{osc.}} \text{ ergs} \quad [\omega_{\text{osc.}} \text{ in c/sec}]$$

or

$$\varepsilon_0 = \tfrac{1}{2}\bar{\omega}_{\text{osc.}} \text{ cm}^{-1} \quad [\bar{\omega}_{\text{osc.}} \text{ in cm}^{-1}]. \tag{3.7}$$

The implication is that the diatomic molecule (and, indeed, *any* molecule) can never have zero vibrational energy; the atoms can never be completely at rest relative to each other.   The quantity

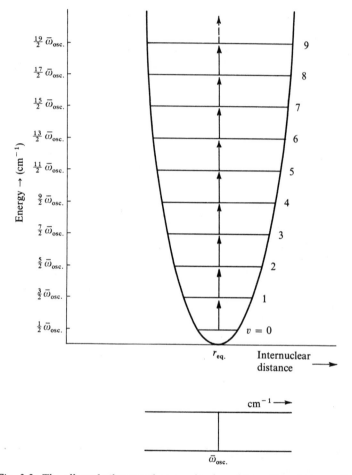

Fig. 3.2: The allowed vibrational energy levels and transitions between them for a diatomic molecule undergoing simple harmonic motion.

$\frac{1}{2}h\omega_{\text{osc.}}$ ergs or $\frac{1}{2}\bar{\omega}_{\text{osc.}}$ cm^{-1} is known as the zero-point energy; it depends only on the classical vibration frequency and hence (equation (3.3) or (3.4)) on the strength of the chemical bond and the atomic masses.

The prediction of zero-point energy is the basic difference between the wave mechanical and classical approaches to molecular vibrations. Classical mechanics could find no objection to a molecule possessing no vibrational energy but wave mechanics insists that it must always vibrate to some extent; the latter conclusion has been amply borne out by experiment.

Further use of the Schrödinger equation leads to the simple *selection rule* for the harmonic oscillator undergoing vibrational changes:

$$\Delta v = \pm 1. \tag{3.8}$$

To this we must, of course, add the condition that vibrational energy changes will only give rise to an observable spectrum if the vibration can interact with radiation, i.e. (cf. Chapter 1) if the vibration involves a change in the dipole moment of the molecule. Thus vibrational spectra will be observable only in heteronuclear diatomic molecules since homonuclear molecules have no dipole moment.

Applying the selection rule we have immediately:

$$\varepsilon_{v+1 \to v} = (v+1+\tfrac{1}{2})\bar{\omega}_{\text{osc.}} - (v+\tfrac{1}{2})\bar{\omega}_{\text{osc.}}$$

$$= \bar{\omega}_{\text{osc.}} \ \text{cm}^{-1} \tag{3.9a}$$

for emission and

$$\varepsilon_{v \to v+1} = \bar{\omega}_{\text{osc.}} \ \text{cm}^{-1} \tag{3.9b}$$

for absorption whatever the initial value of v.

Such a simple result is also obvious from Fig. 3.2—since the vibrational levels are equally spaced, transitions between any two neighbouring states will give rise to the same energy change. Further, since the difference between energy levels expressed in cm^{-1} gives directly the frequency of the spectral line absorbed or emitted,

$$\nu_{\text{spectroscopic}} = \varepsilon = \bar{\omega}_{\text{osc.}} \ \text{cm}^{-1}. \tag{3.10}$$

This, again, is obvious if one considers the mechanism of absorption or emission in classical terms. In absorption, for instance, the vibrating molecule will absorb energy only from radiation with which it can coherently interact (cf. Fig. 1.8) and this must be radiation of its own oscillation frequency.

1.3. The Anharmonic Oscillator. Real molecules do not obey exactly the laws of simple harmonic motion; real bonds, although elastic, are not so homogeneous as to obey Hooke's law. If the bond between atoms is stretched, for instance, there comes a point at which it will break—the molecule dissociates into atoms. Thus although for small compressions and extensions the bond may be taken as perfectly elastic, for larger amplitudes—say greater than 10% of the bond length—a much more complicated behaviour must be assumed. Figure 3.3 shows, diagrammatically, the shape of the energy curve for a typical diatomic molecule, together with (dotted) the ideal, simple harmonic parabola.

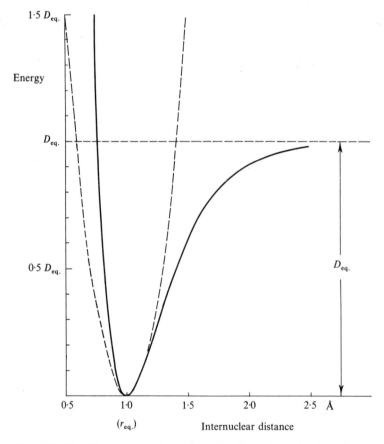

Fig. 3.3: *The Morse curve: the energy of a diatomic molecule undergoing anharmonic extensions and compressions.*

A purely empirical expression which fits this curve to a good approximation was derived by P. M. Morse, and is called the Morse Function:

$$E = D_{eq.}[1-\exp\{a(r_{eq.}-r)\}]^2, \qquad (3.11)$$

where a is a constant for a particular molecule and $D_{eq.}$ is the dissociation energy.

When equation (3.11) is used instead of equation (3.2) in the Schrödinger equation, the pattern of the allowed vibrational energy levels is found to be:

$$\varepsilon_v = (v+\tfrac{1}{2})\bar{\omega}_e - (v+\tfrac{1}{2})^2\bar{\omega}_e x_e \text{ cm}^{-1} \quad (v = 0, 1, 2, \ldots), \quad (3.12)$$

where $\bar{\omega}_e$ is an oscillation frequency (expressed in wavenumbers) which we shall define more closely below, and x_e is the corresponding anharmonicity constant which, for bond stretching vibrations, is always small and positive ($\approx +0.01$), so that the vibrational levels crowd more closely together with increasing v. Some of these levels are sketched in Fig. 3.4.

It should be mentioned that equation (3.12), like (3.11) is an approximation only; more precise expressions for the energy levels require cubic, quartic, etc., terms in $(v+\tfrac{1}{2})$ with anharmonicity constants y_e, z_e, etc., rapidly diminishing in magnitude. These terms are important only at large values of v, and we shall ignore them.

If we rewrite equation (3.12), for the anharmonic oscillator, as:

$$\varepsilon_v = \bar{\omega}_e\{1-x_e(v+\tfrac{1}{2})\}(v+\tfrac{1}{2}) \qquad (3.13)$$

and compare with the energy levels of the *harmonic* oscillator (equation (3.6)), we see that we can write:

$$\bar{\omega}_{osc.} = \bar{\omega}_e\{1-x_e(v+\tfrac{1}{2})\}. \qquad (3.14)$$

Thus the anharmonic oscillator behaves like the harmonic oscillator but with an oscillation frequency which decreases steadily with increasing v. If we now consider the hypothetical energy state obtained by putting $v=-\tfrac{1}{2}$ (at which, according to equation (3.13), $\varepsilon=0$) the molecule would be at the equilibrium point with zero vibrational energy. Its oscillation frequency (in cm^{-1}) would be:

$$\bar{\omega}_{osc.} = \bar{\omega}_e.$$

Thus we see that $\bar{\omega}_e$ may be defined as the (hypothetical) *equilibrium oscillation frequency* of the anharmonic system—the frequency for infinitely small vibrations about the equilibrium point. For any real state specified by a positive integral v the oscillation frequency will be given by equation (3.14). Thus in the ground state ($v=0$)

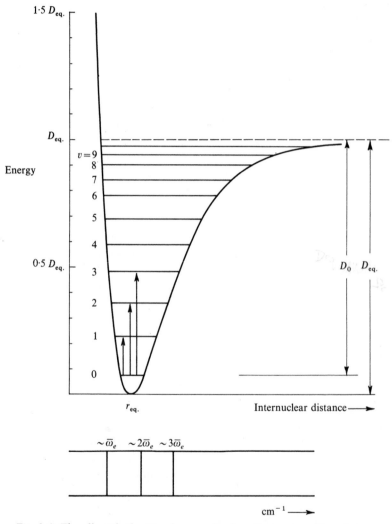

Fig. 3.4: *The allowed vibrational energy levels and some transitions between them for a diatomic molecule undergoing anharmonic oscillations.*

we would have:

$$\bar{\omega}_0 = \bar{\omega}_e(1-\tfrac{1}{2}x_e) \text{ cm}^{-1}$$

and

$$\varepsilon_0 = \tfrac{1}{2}\bar{\omega}_e(1-\tfrac{1}{2}x_e) \text{ cm}^{-1}$$

and we see that the zero point energy differs slightly from that for the harmonic oscillator (equation (3.7)).

The selection rules for the anharmonic oscillator are found to be:

$$\Delta v = \pm 1, \pm 2, \pm 3, \ldots$$

Thus they are the same as for the harmonic oscillator, with the additional possibility of larger jumps. These, however, are predicted by theory and observed in practice to be of rapidly diminishing probability and normally only the lines of $\Delta v = \pm 1, \pm 2$ and ± 3, at the most, have observable intensity. Further at room temperature nearly all the molecules in a particular sample have only the zero-point vibrational energy and exist in the $v=0$ state, because $h\omega_e \gg kT$. In absorption then, we can restrict ourselves to the three possible transitions:

(i) $v=0 \rightarrow v=1$, $\Delta v = +1$, with considerable intensity.

$$\Delta\varepsilon = \varepsilon_{v=1} - \varepsilon_{v=0}$$
$$= (1+\tfrac{1}{2})\bar{\omega}_e - x_e(1+\tfrac{1}{2})^2\bar{\omega}_e - \{\tfrac{1}{2}\bar{\omega}_e - (\tfrac{1}{2})^2 x_e\bar{\omega}_e\}$$
$$= \bar{\omega}_e(1-2x_e) \text{ cm}^{-1}. \tag{3.15a}$$

(ii) $v=0 \rightarrow v=2$, $\Delta v = +2$, with small intensity.

$$\Delta\varepsilon = (2+\tfrac{1}{2})\bar{\omega}_e - x_e(2+\tfrac{1}{2})^2\bar{\omega}_e - \{\tfrac{1}{2}\bar{\omega}_e - (\tfrac{1}{2})^2 x_e\bar{\omega}_e\}$$
$$= 2\bar{\omega}_e(1-3x_e) \text{ cm}^{-1}. \tag{3.15b}$$

(iii) $v=0 \rightarrow v=3$, $\Delta v = +3$, with normally negligible intensity.

$$\Delta\varepsilon = (3+\tfrac{1}{2})\bar{\omega}_e - \{\tfrac{1}{2}\bar{\omega}_e - (\tfrac{1}{2})^2 x_e\bar{\omega}_e\}$$
$$= 3\bar{\omega}_e(1-4x_e) \text{ cm}^{-1}. \tag{3.15c}$$

These three transitions are shown in Fig. 3.4. To a good approximation, since $x_e \approx 0.01$, the three spectral lines lie very close to $\bar{\omega}_e$, $2\bar{\omega}_e$ and $3\bar{\omega}_e$. The line near $\bar{\omega}_e$ is called the *fundamental absorption*, while those near $2\bar{\omega}_e$ and $3\bar{\omega}_e$ are called the *first* and *second overtones*, respectively. The spectrum of HCl, for instance, shows a very intense absorption at 2886 cm^{-1}, a weaker one at 5668 cm^{-1} and a very weak one at 8347 cm^{-1}. If we wish to find the equilibrium frequency of the molecule from these data, we must solve any two of the three equations (cf. equations (3.15)):

$$\bar{\omega}_e(1-2x_e) = 2886$$

$$2\bar{\omega}_e(1-3x_e) = 5668$$

$$3\bar{\omega}_e(1-4x_e) = 8347 \text{ cm}^{-1}$$

and we find $\bar{\omega}_e = 2990$ cm^{-1}, $x_e = 0.0174$. Thus we see that, whereas for the ideal harmonic oscillator the spectral absorption occurred *exactly* at the classical vibration frequency, for real, anharmonic molecules the observed fundamental absorption frequency and the equilibrium frequency may differ considerably.

The force constant of the bond in HCl may be calculated directly from equation (2.22) by inserting the value of $\bar{\omega}_e$:

$$k = 4\pi^2\bar{\omega}_e^2c^2\mu \text{ dynes/cm}$$

$$= 5.16 \times 10^5 \text{ dynes/cm}$$

when the fundamental constants and the reduced mass are inserted. These data, together with that for a few of the very many other diatomic molecules studied by infra-red techniques, are collected in Table 3.1.

TABLE 3.1: *Some Molecular Data for Diatomic Molecules Determined by Infra-Red Spectroscopy*

Molecule	Vibration (cm^{-1})	Anharmonicity Constant x_e	Force Constant (dynes/cm)	Internuclear Distance $r_{eq.}$(Å)
HF	4138.5	0.0218	9.66×10^5	0.927
HCl*	2990.6	0.0174	5.16×10^5	1.274
HBr	2649.7	0.0171	4.12×10^5	1.414
HI	2309.5	0.0172	3.14×10^5	1.609
CO	2169.7	0.0061	19.02×10^5	1.130
NO	1904.0	0.0073	15.95×10^5	1.151
ICl*	384.2	0.0038	2.38×10^5	2.321

* Data refers to the ^{35}Cl isotope.

For completeness we must also consider the possibility of absorption from the $v=1$ state to higher states. As we saw earlier, few molecules normally exist in the $v=1$ state initially at room temperatures, but if the temperature is high (say above 200°C), or the molecule has a particularly low vibration frequency, or the vibration is *very* anharmonic, a sufficient fraction of the molecular population may be in the $v=1$ state to give rise to a weak absorption characterized by the change $v=1 \rightarrow v=2$. We may calculate its wavenumber:

(iv) $v=1 \rightarrow v=2$, $\Delta v = +1$, normally very weak,

$$\Delta \varepsilon = 2\tfrac{1}{2}\bar{\omega}_e - 6\tfrac{1}{4}x_e\bar{\omega}_e - \{1\tfrac{1}{2}\bar{\omega}_e - 2\tfrac{1}{4}x_e\bar{\omega}_e\}$$

$$= \bar{\omega}_e(1-4x_e) \text{ cm}^{-1}. \qquad (3.15d)$$

Thus, should this weak absorption arise, it will be found close to and at slightly *lower* wavenumber than the fundamental (since x_e is small and positive). Such weak absorptions are usually called *hot bands* since a high temperature is one condition for their occurrence. Their nature may be confirmed by raising the temperature of the sample when a true hot band will increase in intensity.

We turn now to consider a diatomic molecule undergoing simultaneous vibration and rotation.

2. The Diatomic Vibrating-Rotator

We saw in Chapter 2 that a typical diatomic molecule has rotational energy separations of 1–10 cm^{-1}, while in the preceding section we found that the vibrational energy separations of HCl were nearly 3000 cm^{-1}. Since the energies of the two motions are so different we may, as a first approximation, consider that a diatomic molecule can execute rotations and vibrations quite independently. This, which we shall call the Born–Oppenheimer approximation (although, cf. equation (6.1), this strictly includes electronic energies), is tantamount to assuming that the combined rotational–vibrational energy is simply the sum of the separate energies:

$$E_{total} = E_{rot.} + E_{vib.} \quad \text{(ergs)}$$

$$\varepsilon_{total} = \varepsilon_{rot.} + \varepsilon_{vib.} \quad \text{(cm}^{-1}\text{)}. \qquad (3.16)$$

We shall see later in what circumstances this approximation does not apply.

Taking the separate expressions for $\varepsilon_{rot.}$ and $\varepsilon_{vib.}$ from equations (2.26) and (3.12) respectively, we have:

$$\varepsilon_{J,v} = \varepsilon_J + \varepsilon_v$$

$$= BJ(J+1) - DJ^2(J+1)^2 + HJ^3(J+1)^3 + \cdots$$

$$+ (v+\tfrac{1}{2})\bar{\omega}_e - x_e(v+\tfrac{1}{2})^2\bar{\omega}_e \text{ cm}^{-1}. \tag{3.17}$$

Initially we shall ignore the small centrifugal distortion constants D, H, etc, and hence write

$$\varepsilon_{total} = \varepsilon_{J,v} = BJ(J+1) + (v+\tfrac{1}{2})\bar{\omega}_e - x_e(v+\tfrac{1}{2})^2\bar{\omega}_e. \tag{3.18}$$

Note, however, that it is not logical to ignore D since this implies that we are treating the molecule as rigid, yet vibrating! The retention of D would have only a very minor effect on the spectrum.

The rotational levels are sketched in Fig. 3.5 for the two lowest

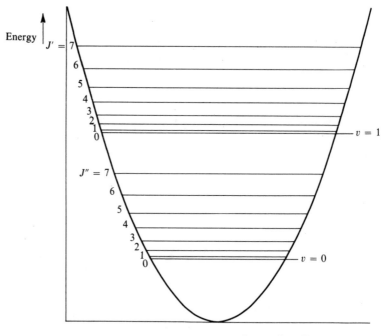

Fig. 3.5: The rotational energy levels for two different vibrational states of a diatomic molecule.

vibrational levels, $v=0$ and $v=1$. There is, however, no attempt at scale in this diagram since the separation between neighbouring J values is, in fact, only some $1/1000$ of that between the v values. Note that since the rotational constant B in equation (3.18) is taken to be the same for all J and v (a consequence of the Born–Oppenheimer assumption), the separation between two levels of given J is the same in the $v=0$ and $v=1$ states.

It may be shown that the selection rules for the combined motions are the same as those for each separately; therefore we have:

$$\Delta v = \pm 1, \pm 2, \text{etc} \qquad \Delta J = \pm 1. \qquad (3.19)$$

Strictly speaking we may also have $\Delta v=0$, but this corresponds to the purely rotational transitions already dealt with in Chapter 2. · Note carefully, however, that a *diatomic* molecule, except under very special and rare circumstances, may *not* have $\Delta J=0$; in other words a vibrational change *must* be accompanied by a simultaneous rotational change.

In Fig. 3.6 we have drawn some of the relevant energy levels and transitions, designating rotational quantum numbers in the $v=0$ state as J'' and in the $v=1$ state as J'. The use of a single prime for the upper state and a double for the lower state is conventional in all branches of spectroscopy.

Remember (and cf. Chapter 2, equation (2.21)) that the rotational levels J'' are filled to varying degrees in any molecular population, so the transitions shown will occur with varying intensities. This is indicated schematically in the spectrum at the foot of Fig. 3.6.

An analytical expression for the spectrum may be obtained by applying the selection rules (equation (3.19)) to the energy levels (equation (3.18)). Considering only the $v=0 \rightarrow v=1$ transition we have in general:

$$\begin{aligned} \Delta\varepsilon_{J,v} &= \varepsilon_{J',v=1} - \varepsilon_{J'',v=0} \\ &= BJ'(J'+1) + 1\tfrac{1}{2}\bar{\omega}_e - 2\tfrac{1}{4}x_e\bar{\omega}_e - \{BJ''(J''+1) + \tfrac{1}{2}\bar{\omega}_e - \tfrac{1}{4}x_e\bar{\omega}_e\} \\ &= \bar{\omega}_o + B(J'-J'')(J'+J''+1) \text{ cm}^{-1}, \end{aligned}$$

where, for brevity, we write $\bar{\omega}_o$ for $\bar{\omega}_e(1-2x_e)$.

We should note that taking B to be identical in the upper and lower vibrational states is a direct consequence of the Born–Oppenheimer approximation—rotation is unaffected by vibrational changes.

6

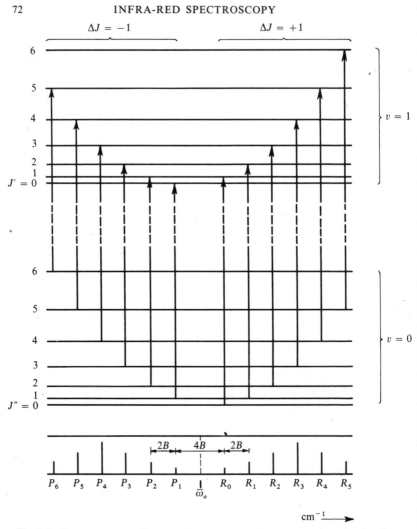

Fig. 3.6: *Some transitions between the rotational–vibrational energy levels of a diatomic molecule together with the spectrum arising from them.*

Now we can have:

(*i*) $\Delta J = +1$, i.e. $J' = J'' + 1$ or $J' - J'' = +1$; hence

$$\Delta\varepsilon_{J,v} = \bar{\omega}_o + 2B(J''+1) \text{ cm}^{-1}, \quad J'' = 0, 1, 2, \ldots \quad (3.20a)$$

(*ii*) $\Delta J = -1$, i.e. $J'' = J' + 1$ or $J' - J'' = -1$; and

$$\Delta\varepsilon_{J,v} = \bar{\omega}_o - 2B(J'+1) \text{ cm}^{-1}, \quad J' = 0, 1, 2, \ldots \quad (3.20b)$$

These two expressions may conveniently be combined into:

$$\Delta\varepsilon_{J,v} = \bar{v}_{\text{spec.}} = \bar{\omega}_o + 2Bm \text{ cm}^{-1}, \quad m = \pm1, \pm2,\dots, \quad (3.20c)$$

where m, replacing $J''+1$ in equation (3.20a) and $J'+1$ in equation (3.20b) has positive values for $\Delta J = +1$ and is negative if $\Delta J = -1$. Note particularly that m *cannot be zero* since this would imply values of J' or J'' to be -1. The frequency $\bar{\omega}_o$ is usually called the *band origin* or *band centre*.

Equation (3.20c), then, represents the combined vibration–rotation spectrum. Evidently it will consist of equally spaced lines (spacing $= 2B$) on each side of the band origin $\bar{\omega}_o$, but, since $m \neq 0$, the *line at $\bar{\omega}_o$ itself will not appear*. Lines to the low frequency side of $\bar{\omega}_o$, corresponding to negative m (i.e. $\Delta J = -1$) are referred to as the *P branch*, while those to the high frequency side (m positive, $\Delta J = +1$) are called the *R branch*. This apparently arbitrary notation may become clearer if we state here that later, in other contexts, we shall be concerned with ΔJ values of 0 and ±2, in addition to ±1 considered here; the labelling of line series is then quite consistent:

lines arising from $\Delta J =$	-2	-1	0	$+1$	$+2$	
called:		O	P	Q	R	S branch.

The P and R notation, with the *lower J* (J'') value as a suffix, is illustrated on the diagrammatic spectrum of Fig. 3.6. This is the conventional notation for such spectra.

It is readily shown that the inclusion of the centrifugal distortion constant D leads to the following expression for the spectrum:

$$\Delta\varepsilon = \bar{v}_{\text{spect.}} = \bar{\omega}_o + 2Bm - 4Dm^3 \text{ cm}^{-1} \quad (m = \pm1, \pm2, \pm3,\dots).$$
$$(3.21)$$

But we have seen in Chapter 2 that B is some 10 cm^{-1} or less, while D is only some 0·01% of B. Since a good infra-red spectrometer has a resolving power of about 0·5 cm^{-1} it is obvious that D is negligible to a very high degree of accuracy.

The anharmonicity factor, on the other hand, is not negligible. It affects not only the position of the band origin (since $\bar{\omega}_o = \bar{\omega}_e(1 - 2x_e)$), but, by extending the selection rules to include $\Delta v = \pm2, \pm3$, etc., also allows the appearance of overtone bands having identical rotational structure. This is illustrated in Fig. 3.7(a), where the fundamental and first overtone of carbon monoxide are shown.

From the band centres we can calculate, as shown in Section 1.3, the equilibrium frequency $\bar{\omega}_e$ and the anharmonicity constant x_e.

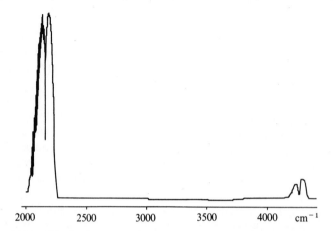

Fig. 3.7(a): *The fundamental absorption (centred at about 2143 cm^{-1}) and the first overtone (centred at about 4260 cm^{-1}) of carbon monoxide; the fine structure of the P branch in the fundamental is partially resolved. (Gas pressure 650 mm* Hg *in a 10 cm cell.)*

3. The Vibration–Rotation Spectrum of Carbon Monoxide

In Fig. 3.7(*b*) we show the fundamental vibration–rotation band of carbon monoxide under high resolution, with some lines in the *P* and *R* branches numbered according to their *J''* values. Table 3.2 gives the observed wavenumbers of the first five lines in each branch.

TABLE 3.2: Part of the Infra-Red Spectrum of Carbon Monoxide

Line	$\bar{\nu}$	Separation $\Delta\bar{\nu}$	Line	$\bar{\nu}$	Separation $\Delta\bar{\nu}$
$P_{(1)}$	2139·43		$R_{(0)}$	2147·08	
		3·88			3·78
$P_{(2)}$	2135·55		$R_{(1)}$	2150·86	
		3·92			3·73
$P_{(3)}$	2131·63		$R_{(2)}$	2154·59	
		3·95			3·72
$P_{(4)}$	2127·68		$R_{(3)}$	2158·31	
		3·98			3·66
$P_{(5)}$	2123·70		$R_{(4)}$	2161·97	

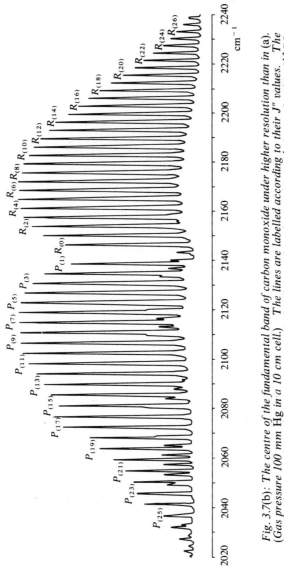

Fig. 3.7(b): *The centre of the fundamental band of carbon monoxide under higher resolution than in (a). (Gas pressure 100 mm Hg in a 10 cm cell.) The lines are labelled according to their J″ values. The P branch is complicated by the presence of a band centred at about 2100 cm⁻¹ due to the 1% of ¹³CO in the sample; some of the rotational lines from this band appear between P branch lines, others are overlapped by a P branch line and give it an enhanced intensity (e.g. lines $P_{(16)}$, $P_{(17)}$, $P_{(23)}$ and $P_{(24)}$).*

We shall discuss shortly the slight decrease in separation between the rotational lines as the wavenumber increases; this decrease is apparent from the table and from a close inspection of the "wings" of the spectrum.

From the table we see that the band centre is at about 2143 cm^{-1} while the average line separation near the centre is 3·83 cm^{-1}. This immediately gives:

$$2B = 3·83 \text{ cm}^{-1}, \qquad B = 1·915 \text{ cm}^{-1}.$$

This is in satisfactory agreement with the value $B = 1·92118$ cm^{-1} derived by microwave studies (cf. Chapter 2, Section 3.1) and we could, therefore, have obtained quite good values for the rotational constant and hence the moment of inertia and bond length from infra-red data alone. Historically, of course, the infra-red values came first, the more precise microwave values following much later.

It is worth noting at this point that approximate rotational data is obtainable from spectra even if the separate rotational lines are not resolved. Thus Fig. 3.8 shows the spectrum of carbon monoxide under much poorer resolution, when the rotational fine structure is blurred out to an envelope. Now we saw in Chapter 2, equation (2.21a), that the maximum population of levels, and hence maximum intensity of transition, occurs at a J value of $\sqrt{kT/2Bhc} - \frac{1}{2}$. Remembering that $m = J + 1$ we substitute in equation (3.20c) $m = \pm\sqrt{kT/2Bhc} + \frac{1}{2}$ and obtain:

$$R \text{ branch max.:} \qquad \bar{v}_{\text{max.}} = \bar{\omega}_o + 2B\left(\sqrt{\frac{kT}{2Bhc}} + \frac{1}{2}\right)$$

$$P \text{ branch max.:} \qquad \bar{v}_{\text{max.}} = \bar{\omega}_o - 2B\left(\sqrt{\frac{kT}{2Bhc}} + \frac{1}{2}\right).$$

The *separation* between the two maxima, $\Delta\bar{v}$, is then:

$$\Delta\bar{v} = 4B\left(\sqrt{\frac{kT}{2Bhc}} + \frac{1}{2}\right) = \sqrt{\frac{8kTB}{hc}} + 2B.$$

In the case of carbon monoxide we see that the separation is about 50 cm^{-1}, while the temperature at which the spectrum was obtained was about 300° Abs. We are led, then, to a B value of about 1·7 cm^{-1} which is in fair agreement with the earlier values, but much less precise.

From Table 3.2 we see that the band origin, at the midpoint of $P_{(1)}$ and $R_{(0)}$, is at 2143·26 cm^{-1}. This, then, is the fundamental vibration frequency of carbon monoxide, if anharmonicity is

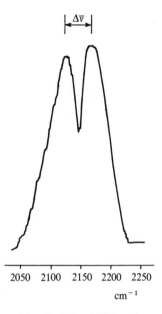

$$2050 \quad 2100 \quad 2150 \quad 2200 \quad 2250$$

$$\text{cm}^{-1}$$

Fig. 3.8: The fundamental band of Fig. 3.7(b) under very low resolution; all rotational fine structure has been lost and a typical PR contour is seen.

ignored. The latter can be taken into account, however, since the first overtone is found to have its origin at 4260·04 cm^{-1}. We have:

$$\bar{\omega}_e(1 - 2x_e) = \bar{\omega}_o = 2143\cdot26$$

$$2\bar{\omega}_e(1 - 3x_e) = 4260\cdot04$$

from which $\bar{\omega}_e = 2169\cdot74$ cm^{-1}, $x_e = 0\cdot0061$.

4. Breakdown of the Born–Oppenheimer Approximation: the Interaction of Rotations and Vibrations

So far we have assumed that vibration and rotation can proceed quite independently of each other. A molecule vibrates some 10^3 times during the course of a single rotation, however, so it is evident that the bond length (and hence the moment of inertia and B constant) also changes continually during the rotation. If the vibration is simple-harmonic the mean bond length will be the same as the equilibrium bond length and it will not vary with vibrational energy; this is seen in Fig. 3.1. However, the rotational constant B depends on $1/r^2$ and, as shown by an example in Chapter 2, Section 3.4, the average value of this quantity is not the same as $1/r^2_{eq.}$ where $r_{eq.}$ is the equilibrium length. Further an increase in the vibrational energy is accompanied by an increase in the vibrational amplitude and hence the value of B will depend on the v quantum number.

In the case of anharmonic vibrations the situation is rather more complex. Now an increase in vibrational energy will lead to an increase in the average bond length—this is perhaps most evident from Fig. 3.4. The rotational constant then varies even more with vibrational energy.

In general, it is plain that, since $r_{av.}$ *increases* with the vibrational energy, B is *smaller* in the upper vibrational state than in the lower. In fact an equation of the form:

$$B_v = B_e - \alpha(v + \tfrac{1}{2})$$
(3.22)

gives, to a high degree of approximation, the value of B_v, the rotational constant in vibrational level v, in terms of the equilibrium value B_e and α, a small positive constant for each molecule.

Here we restrict our discussion to the fundamental vibrational change, i.e. the change $v=0 \rightarrow v=1$, and we may take the respective B values as B_0 and B_1 with $B_0 > B_1$. For this transition:

$$\Delta\varepsilon = \varepsilon_{J', v=1} - \varepsilon_{J'', v=0}$$
$$= \bar{\omega}_o + B_1 J'(J'+1) - B_0 J''(J''+1) \text{ cm}^{-1},$$

where, as before, $\bar{\omega}_o = \bar{\omega}_e(1 - 2x_e)$.

We then have the two cases:

(i) $\Delta J = +1, \quad J' = J''+1$

$$\Delta\varepsilon = \bar{v}_R = \bar{\omega}_o + (B_1 + B_0)(J''+1) + (B_1 - B_0)(J''+1)^2 \text{ cm}^{-1}$$
($J'' = 0, 1, 2, \ldots$)   (3.23a)

and

(ii) $\qquad \Delta J = -1, \qquad J'' = J'+1$

$$\Delta \varepsilon = \bar{v}_P = \bar{\omega}_o - (B_1 + B_0)(J'+1) + (B_1 - B_0)(J'+1)^2 \ \mathrm{cm}^{-1}$$

$$(J' = 0, 1, 2, \ldots) \quad (3.23b)$$

where we have written $\bar{v}_P$ and $\bar{v}_R$ to represent the wavenumbers of the P and R branch lines respectively. These two equations can be combined into the expression:

$$\bar{v}_{P,R} = \bar{\omega}_o + (B_1 + B_0)m + (B_1 - B_0)m^2 \ \mathrm{cm}^{-1} \quad (m = \pm 1, \pm 2, \ldots)$$

$$(3.23c)$$

where positive m values refer to the R branch and negative to P.

We see that ignoring vibration–rotation interaction involves setting $B_1 = B_0$, when equation (3.23c) immediately simplifies to (3.20c). Since $B_1 < B_0$ the last term of (3.23c) is always negative, irrespective of the sign of m, and the effect on the spectrum of a diatomic molecule is to crowd the rotational lines more closely together with increasing m on the R branch side, while the P branch lines become more widely spaced as (negative) m increases. Normally B_1 and B_0 differ only slightly and the effect is marked only for high m values. This is exactly the situation shown in the spectrum of carbon monoxide, Fig. 3.7(b).

TABLE 3.3: *Observed and Calculated Wavenumbers of some Lines in the Spectrum of Carbon Monoxide*

m	J''	$\bar{v}_{\mathrm{obs.}}$	$\bar{v}_{\mathrm{calc.}}$*
30	29	2241·64	2241·91
25	24	2227·63	2227·65
20	19	2212·62	2212·54
15	14	2196·66	2196·53
10	9	2179·77	2179·66
5	4	2161·97	2161·90
0	—	(Band centre)	2143·28
−5	5	2123·70	2123·78
−10	10	2103·27	2103·40
−15	15	2082·01	2082·15
−20	20	2059·91	2060·02
−25	25	2037·03	2037·02
−30	30	2013·35	2013·14

* Values calculated from: $\bar{v} = 2143 \cdot 28 + 3 \cdot 813m - 0 \cdot 0175m^2$.

In Table 3.3 some of the data for carbon monoxide are tabulated, together with the positions of lines calculated from the equation:

$$\bar{v}_{\text{spect.}} = 2143 \cdot 28 + 3 \cdot 813 m - 0 \cdot 0175 m^2 \text{ cm}^{-1}.$$

From this we see that, for this molecule:

$$B_1 = 1 \cdot 898 \text{ cm}^{-1}, \qquad B_0 = 1 \cdot 915 \text{ cm}^{-1},$$

and hence, using equation (3.22), we have:

$$\alpha = 0 \cdot 018, \qquad B_e = 1 \cdot 924 \text{ cm}^{-1}.$$

Further, we can calculate the equilibrium bond length and the bond lengths in the $v=0$ and $v=1$ states (cf. pp. 31–32) to be:

$$r_{\text{eq.}} = 1 \cdot 130 \text{ Å}, \qquad r_0 = 1 \cdot 133 \text{ Å}, \qquad r_1 = 1 \cdot 136 \text{ Å}.$$

5. The Vibrations of Polyatomic Molecules

In this section and the next, just as in the corresponding one dealing with the pure rotational spectra of polyatomic molecules, we shall find that although there is an increase in the complexity, only slight and quite logical extensions to the simple theory are adequate to give us an understanding of the spectra. We shall need to discuss:

1. The number of fundamental vibrations and their symmetry;
2. The possibility of overtone and combination bands;
3. The influence of rotation on the spectra.

5.1. Fundamental Vibrations and their Symmetry. Consider a molecule containing N atoms: we can refer to the position of each atom by specifying three co-ordinates (e.g. the x, y and z cartesian co-ordinates). Thus the total number of co-ordinate values is $3N$ and we say the molecule has $3N$ *degrees of freedom* since each co-ordinate value may be specified quite independently of the others. However, once all $3N$ co-ordinates have been fixed, the bond distances and bond angles of the molecule are also fixed and no further arbitrary specifications can be made.

Now the molecule is free to move in three-dimensional space, as a whole, without change of shape. We can refer to such movement by noting the position of its centre of gravity at any instant—to do this requires a statement of three co-ordinate values. This translational movement uses three of the $3N$ degrees of freedom leaving $3N - 3$. In general, also, the rotation of a non-linear molecule can be resolved into components about three perpendicular axes (cf. Chapter 2, Section 1). Specification of these axes also requires three degrees of freedom, and the molecule is left with $3N-6$ degrees of freedom. The only other motion allowed to it is internal vibration, so we know immediately that a non-linear N-atomic molecule can have $3N-6$ different internal vibrations:

$$\text{non-linear:} \quad 3N-6 \text{ fundamental vibrations.} \quad (3.24a)$$

If, on the other hand, the molecule is linear, we saw in Chapter 2 that there is no rotation about the bond axis; hence only two degrees of rotational freedom are required, leaving $3N-5$ degrees of vibrational freedom—one more than in the case of a non-linear molecule:

$$\text{linear:} \quad 3N-5 \text{ fundamental vibrations.} \quad (3.24b)$$

In both cases, since an N-atomic molecule has $N-1$ bonds (for acyclic molecules) between its atoms, $N-1$ of the vibrations are bond-stretching motions, the other $2N-5$ (non-linear) or $2N-4$ (linear) are bending motions.

Let us look briefly at examples of these rules. First, we see that for a diatomic molecule (perforce linear) such as we have already considered in this chapter: $N=2$, $3N-5=1$ and thus there can be only one fundamental vibration. Note, however, that the $3N-5$ rule says nothing about the presence, absence or intensity of overtone vibrations—these are governed solely by anharmonicity.

Next, consider water, H_2O. This (Fig. 3.9) is non-linear and triatomic. Also in the figure are the $3N-6=3$ allowed vibrational modes, the arrows attached to each atom showing the direction of its motion during half of the vibration. Each motion is described as stretching or bending depending on the nature of the change in molecular shape.

These three vibrational motions are also referred to as the *normal modes of vibration* (or *normal vibrations*) of the molecule; in general a normal vibration is defined as a molecular motion in which all the atoms move in phase and with the same frequency.

Further each motion of Fig. 3.9 is labelled either symmetric or antisymmetric. It is not necessary here to go far into the matter of general molecular symmetry since other excellent texts already exist for the interested student, but we can see quite readily that the water molecule contains some elements of symmetry. In particular

82 INFRA-RED SPECTROSCOPY

consider the dotted line at the top of Fig. 3.9 which bisects the HOH angle; if we rotate the molecule about this axis by 180° its final appearance is identical with the initial one. This axis is thus referred to as a C_2 axis since twice in every complete revolution the molecule presents an identical aspect to an observer. This particular molecule has only the one rotational symmetry axis, and it is conventional to refer the molecular vibrations to this axis. Thus consider the first vibration, Fig. 3.9(a). If we rotate the *vibrating* molecule by 180° the vibration is quite unchanged in

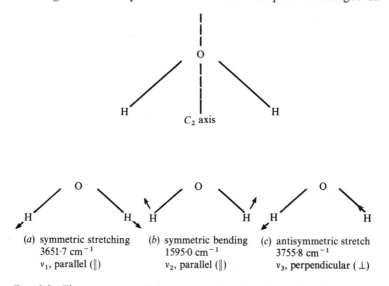

(a) symmetric stretching
3651·7 cm⁻¹
v_1, parallel (∥)

(b) symmetric bending
1595·0 cm⁻¹
v_2, parallel (∥)

(c) antisymmetric stretch
3755·8 cm⁻¹
v_3, perpendicular (⊥)

Fig. 3.9: The symmetry of the water molecule and its three fundamental vibrations. The motion of the oxygen atom, which must occur to keep the centre of gravity of the molecule stationary, is here ignored.

character—we call this a symmetric vibration. The bending vibration, v_2, is also symmetric. Rotation of the stretching motion of Fig. 3.9(c) about the C_2 axis, however, produces a vibration which is in antiphase with the original and so this motion is described as the antisymmetric stretching mode.

In order to be infra-red active, as we have seen, there must be a dipole change during the vibration and this change may take place either along the line of the symmetry axis (parallel to it, or ∥) or at right angles to the line (perpendicular, ⊥). Figure 3.10 shows the nature of the dipole change for the three vibrations of water, and

justifies the labels parallel or perpendicular attached to them in
Fig. 3.9. We shall see later that the distinction is important when
considering the influence of *rotation* on the spectrum.

Finally the vibrations are labelled in Fig. 3.9 as v_1, v_2 and v_3.
By convention it is usual to label vibrations in decreasing frequency

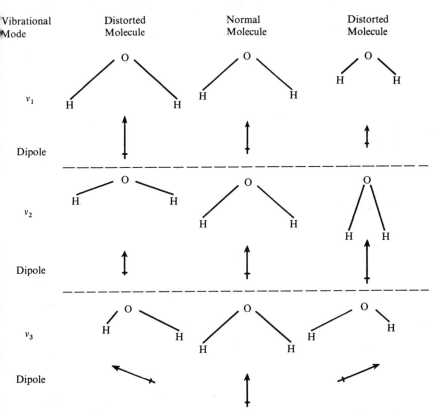

*Fig. 3.10: The change in the electric dipole moment produced by each
vibration of the water molecule; this is seen to occur either along ($\|$) or across
($\perp$) the symmetry axis.   The amplitudes are greatly exaggerated for clarity.*

within their symmetry type. Thus the symmetric vibrations of
H_2O are labelled v_1 for the highest fully symmetric frequency
(3651.7 cm^{-1}), and v_2 for the next highest (1595.0 cm^{-1}); the
antisymmetric vibration at 3755.8 cm^{-1} is then labelled v_3.

Our final example is of the linear triatomic molecule CO_2, for which the normal vibrations are shown in Fig. 3.11. For this molecule there are two different sets of symmetry axes. There is an infinite number of 2-fold axes (C_2) passing through the carbon atom at right angles to the bond direction, and there is an ∞-fold axis (C_∞) passing through the bond axis itself (this is referred to as ∞-fold since rotation of the molecule about the bond axis through *any* angle gives an identical aspect). The names symmetric stretch and antisymmetric stretch are self-evident, but it should be noted

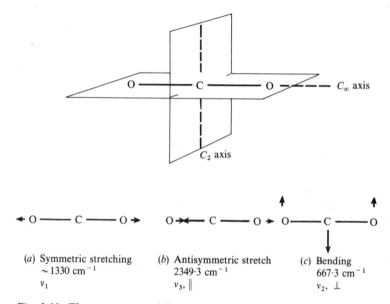

(a) Symmetric stretching
   ~1330 cm^{-1}
   v_1

(b) Antisymmetric stretch
   2349·3 cm^{-1}
   v_3, ∥

(c) Bending
   667·3 cm^{-1}
   v_2, ⊥

Fig. 3.11: The symmetry and fundamental vibrations of the carbon dioxide molecule.

that the symmetric stretch produces no change in the dipole moment (which remains zero) so that this vibration is not infra-red active, and the motion cannot be regarded as ∥ or ⊥ to the ∞-fold symmetry axis. The vibration frequency may be obtained in other ways, however, which we shall discuss in the next chapter.

For linear triatomic molecules, $3N-5=4$, and we would expect four vibrational modes instead of the three shown in Fig. 3.11. However, consideration shows that v_2 in fact consists of *two* vibrations—one in the plane of the paper as drawn, and the other in which the oxygen atoms move simultaneously into and out of the plane. The two sorts of motion are, of course, identical in all respects except direction and are termed *degenerate*; they must, nevertheless, be considered as separate motions, and it is always in the degeneracy of a bending mode that the extra vibration of a linear molecule over a non-linear one is to be found.

It might be thought that v_2 of H_2O (Fig. 3.9(*b*)) could occur by the hydrogens moving simultaneously in and out of the plane of the paper. Such a motion is not a vibration, however, but a *rotation*. As the molecule approaches linearity this rotation degenerates into a vibration, and the molecule loses one degree of rotational freedom in exchange for one of vibration.

5.2. Overtone and Combination Frequencies. If one were able to observe the molecules of H_2O or CO_2 directly their overall vibrations would appear extremely complex; in particular, each atom would not follow tidily any one of the separate paths depicted in Figs. 3.9 or 3.11, but its motion would essentially be a superposition of all such paths, since every possible vibration is always excited, at least to the extent of its zero point energy. However, such superposition could be resolved into its components if, for instance, we could examine the molecules under stroboscopic light flashing at each fundamental frequency in turn. This is, so to speak, the essence of infra-red spectroscopy—instead of flashing we have the radiation frequency, and the "examination" is a sensing of dipole alteration. Thus, as we would expect, the infra-red spectrum of a complex molecule consists essentially of an absorption band at each of the $3N - 6$ (non-linear) or $3N - 5$ (linear) fundamental frequencies.

This is, of course, an oversimplification, in which two approximations are implicit: (*i*) that each vibration is simple harmonic, (*ii*) that each vibration is quite independent and unaffected by the others. We shall consider (*ii*) in more detail later; for the moment we can accept it as a good working approximation.

When the restriction to simple harmonic motion is lifted we have again, as in the case of the diatomic molecule (Section 1.3), the possibility of first, second, etc., overtones occurring at frequencies near $2v_1, 3v_1, \ldots, 2v_2, 3v_2, \ldots, 2v_3, \ldots$, etc., where each v_i is a fundamental mode. The intensities fall off rapidly. However, in addition, the selection rules now permit *combination bands* and *difference bands*. The former arise simply from the addition of two or more fundamental frequencies or overtones. Such combinations as $v_1 + v_2, 2v_1 + v_2, v_1 + v_2 + v_3$, etc., become allowed, although their intensities are normally very small.   Similarly the difference bands, e.g. $v_1 - v_2$, $2v_1 - v_2$, $v_1 + v_2 - v_3$ have small intensities but are often to be found in a complex spectrum.

The intensities of overtone or combination bands may sometimes be considerably enhanced by a resonance phenomenon.   It may happen that two vibrational modes in a particular molecule have frequencies very close to each other—they are described as *accidentally* degenerate.   Note that we are not here referring to identical vibrations, such as the two identical v_2's of CO_2 (Fig. 3.11), but rather to the possibility of two quite different modes having similar energies.   Normally the fundamental modes are quite different from each other and accidental degeneracy is found most often between a fundamental and some overtone or combination.   A simple example is to be found in CO_2 where v_1, described as at about 1330 cm^{-1} is very close to that of $2v_2 = 1334$ cm^{-1}.   (As mentioned earlier, these bands are not observable in the infra-red, but both may be seen in the Raman spectrum discussed in the next chapter; the principles of resonance apply equally to both techniques.)   Quantum mechanics shows that two such bands may interfere with each other in such a way that the higher is raised in frequency, the lower depressed—and in fact the Raman spectrum shows two bands, one at 1285 cm^{-1}, the other at 1385 cm^{-1}. Their mean is plainly at about 1330 cm^{-1}.

Note, however, that one of these bands arises from a fundamental mode (v_1), the other from the overtone $2v_2$, and we would normally expect the former to be much more intense than the latter.   In fact, they are found to be of about the same intensity—the overtone has gained intensity at the expense of the fundamental.   This is an extreme case—normally the overtone takes only a small part of the intensity from the fundamental.   The situation is often likened to that of two pendulums connected to a common bar—when the pendulums have quite different frequencies they oscillate independently; when their frequencies are similar they can readily exchange energy, one with the other, and an oscillation given to one is transferred to and fro between them.   They are said to resonate. Similarly two close molecular vibrational frequencies resonate and exchange energy—the phenomenon being known as *Fermi resonance* when a fundamental resonates with an overtone.   In the spectrum of a complex molecule exhibiting many fundamentals and overtones, there is a good chance of accidental degeneracy, and

Fermi resonance, occurring. However, it should be mentioned that not all such degeneracies lead to resonance. It is necessary, also, to consider the molecular symmetry and the type of degenerate vibrations; we shall not, however, pursue the topic further here.

6. The Influence of Rotation on the Spectra of Polyatomic Molecules

In Section 2 of this chapter we found that the selection rule for the simultaneous rotation and vibration of a diatomic molecule was

$$\Delta v = \pm 1, \qquad \Delta J = \pm 1, \qquad \Delta J \neq 0$$

and that this gave rise to a spectrum consisting of approximately equally spaced line series on each side of a central minimum designated as the band centre.

Earlier in the present section we showed that the vibrations of complex molecules could be subdivided into those causing a dipole change either (*i*) parallel or (*ii*) perpendicular to the major axis of rotational symmetry. The purpose of this distinction, and the reason for repeating it here, is that the selection rules for the *rotational* transitions of complex molecules depend, rather surprisingly, on the type of *vibration*, ∥ or ⊥, which the molecule is undergoing. Less surprisingly, the selection rules and the energies, depend on the *shape* of the molecule also. We shall deal first with the linear molecule as the simplest, and then say a few words about the other types of molecule.

6.1. Linear Molecules. (*i*) Parallel Vibrations. The selection rule for these is identical with that for diatomic molecules, i.e.

$$\Delta J = \pm 1, \quad \Delta v = \pm 1 \quad \text{for simple harmonic motion} \qquad (3.25a)$$

$$\Delta J = \pm 1, \quad \Delta v = \pm 1, \pm 2, \pm 3, \dots \quad \text{for anharmonic motion.}$$

$$(3.25b)$$

(This is, in fact, as expected, since a diatomic molecule is linear and can undergo only parallel vibrations.) The spectra will thus be similar in appearance, consisting of P and R branches with lines about equally spaced on each side, no line occurring at the band centre. Now, however, the moment of inertia may be considerably larger, the B value correspondingly smaller, and the P or R line spacing will be less. Figure 3.12 shows part of the spectrum of HCN, a linear molecule whose structure is H—C≡N. The band

7

concerned is the symmetric stretching frequency at about 3310 cm^{-1}, (corresponding to the v_1 mode of CO_2 in Fig. 3.11), and the spacing is observed to be about 2·8–3·0 cm^{-1} near the band centre. This is to be compared, for example, with the spacing of about 4·0 cm^{-1} in the case of CO.

For still larger molecules the value of B may be so small that separate lines can no longer be resolved in the P and R branches.

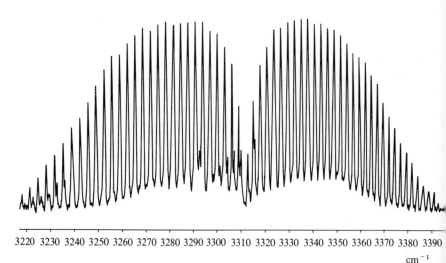

3220 3230 3240 3250 3260 3270 3280 3290 3300 3310 3320 3330 3340 3350 3360 3370 3380 3390

cm^{-1}

Fig. 3.12: Spectrum of the symmetric stretching vibration of the HCN molecule showing the P and R branch lines.

In this case the situation is exactly analogous to that shown previously in Fig. 3.8 and the same remarks apply as to the possibility of deriving a rough value of B from the separation between the maxima of the P and R envelopes. We shall shortly see that a non-linear molecule cannot give rise to this type of band shape, so its observation somewhere within a spectrum is sufficient proof that a linear, or nearly linear, molecule is being studied.

(*ii*) Perpendicular Vibrations. For these the selection rule is found to be:

$$\Delta v = \pm 1, \quad \Delta J = 0, \pm 1 \quad \text{for simple harmonic motion} \quad (3.26)$$

which implies that now, for the first time, a vibrational change can

take place with *no simultaneous rotational transition*.   The result
is illustrated in Fig. 3.13, which shows the same energy levels and
transitions as Fig. 3.6 with the addition of $\Delta J=0$ transitions.   If
the oscillation is taken as simple harmonic the energy levels are
identical with those of equation (3.18) and the P and R branch lines

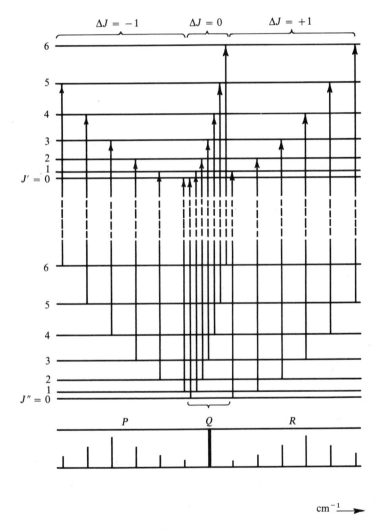

*Fig. 3.13: The rotational energy levels for two vibrational states showing the
effect on the spectrum of transitions for which $\Delta J=0$.*

are given, as before, by equations (3.20) or (3.21). Transitions with $\Delta J = 0$, however, correspond to a Q *branch* whose lines may be derived from the equations:

$$\Delta\varepsilon = \varepsilon_{J,\,v+1} - \varepsilon_{J,\,v}$$

$$= 1\tfrac{1}{2}\bar{\omega}_e - 2\tfrac{1}{4}x_e\bar{\omega}_e + BJ(J+1) - \{\tfrac{1}{2}\bar{\omega}_e - \tfrac{1}{4}x_e\bar{\omega}_e + BJ(J+1)\}$$

$$= \bar{\omega}_o \text{ cm}^{-1}, \quad \text{for all } J. \tag{3.27}$$

Thus the Q branch consists of lines superimposed upon each other at the band centre $\bar{\omega}_o$, one contribution arising for each of the populated J values. The resultant line is usually very intense.

If we take into account the fact that the B values differ slightly in the upper and lower vibrational states (cf. Section 4), we would write instead:

$$\Delta\varepsilon = \varepsilon_{J,\,v+1} - \varepsilon_{J,\,v}$$

$$= 1\tfrac{1}{2}\bar{\omega}_e - 2\tfrac{1}{4}x_e\bar{\omega}_e + B'J(J+1) - \{\tfrac{1}{2}\bar{\omega}_e - \tfrac{1}{4}x_e\bar{\omega}_e + B''J(J+1)\}$$

$$= \bar{\omega}_o + J(J+1)(B' - B''). \tag{3.28}$$

Further, if $B' < B''$, we see that the Q branch line would become split into a series of lines on the *low* frequency side of $\bar{\omega}_o$ (since $B' - B''$ is negative). Normally, however, $B' - B''$ is so small that the lines cannot be resolved, and the Q branch appears as a somewhat broad resonance centred around $\bar{\omega}_o$. This is illustrated in Fig. 3.14, which is a spectrum of the bending mode of HCN (corresponding to v_2 of CO_2 in Fig. 3.11). Finally, if the rotational fine structure is unresolved, this type of band has the distinctive contour shown in Fig. 3.15.

It should be remembered (see Chapter 2) that polyatomic molecules with zero dipole moment do not give rise to pure rotation spectra in the microwave region (e.g. CO_2, HC≡CH, CH_4). Such molecules do, however, show vibrational spectra in the infra-red region (or Raman, cf. Chapter 4) and, if these spectra exhibit resolved fine structure, the moment of inertia of the molecule can be obtained.

6.2. The Influence of Nuclear Spin. It is necessary here to say a brief word about the spectrum of carbon dioxide and other linear molecules possessing a centre of symmetry. A centre of symmetry means that identical atoms are symmetrically disposed with respect to the centre of gravity of the molecule. Thus, plainly both CO_2

[O=C=O], and acetylene [H—C≡C—H] possess a centre of symmetry, while HCN, or N_2O [N≡N=O] do not.

The reader may have noticed that, although we used CO_2 as an example of a vibrating molecule in Fig. 3.11, we did not use it to illustrate real spectra in the subsequent discussion. This is because the centre of symmetry has an effect on the intensity of alternate lines in the P and R branches. The effect is due to the

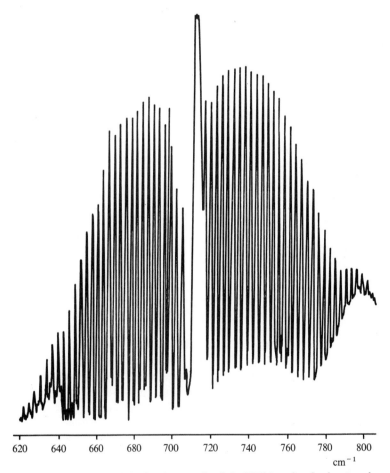

Fig. 3.14: Spectrum of the bending mode of the HCN molecule showing the PQR structure. The broad absorption centred at 800 cm^{-1} is due to an impurity.

existence of nuclear spin (cf. Chapter 7) and is an additional factor determining the populations of rotational levels. In the case of CO_2 every alternate rotational level is completely unoccupied and so alternate lines in the P and R branches have zero intensity. This leads to a line spacing of $4B$ instead of the usual $2B$ discussed above. That the spacing is indeed $4B$ (and not $2B$ with an un-expectedly large value of B) can be shown in several ways, perhaps

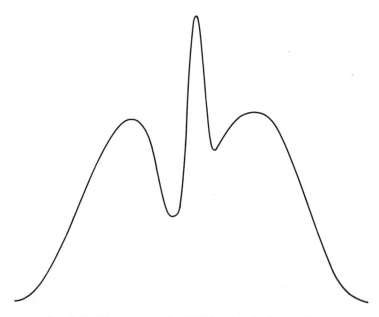

Fig. 3.15: The contour of a PQR band under low resolution.

the most convincing of which is to examine the spectrum of the isotopic molecule ^{18}O—C—^{16}O. Here there is no longer a centre of symmetry, nuclear spin does not now affect the spectrum and the line spacing is found to be just half that for "normal" CO_2.

In the case of acetylene, alternate levels have populations which differ by a factor of 3:1 (this, due to nuclear spin alone, is super-imposed on the normal thermal distribution and degeneracy) so that the P and R branch lines show a strong, weak, strong, weak,... al-ternation in intensity, as shown in Fig. 3.16.

6.3. Symmetric Top Molecules. Following the Born–Oppen-heimer approximation we can take the vibrational–rotational energy

levels for this type of molecule to be the sum of the vibrational
levels:

$$\varepsilon_{\text{vib.}} = (v+\tfrac{1}{2})\bar{\omega}_e - (v+\tfrac{1}{2})^2 x_e\bar{\omega}_e \text{ cm}^{-1}, \qquad [v = 0, 1, 2, 3, \ldots]$$

and the rotational levels (cf. Chapter 2, equation (2.38)),

$$\varepsilon_{\text{rot.}} = BJ(J+1) + (A-B)K^2 \text{ cm}^{-1}, \quad [J = 0, 1, 2, \ldots;$$
$$K = J, (J-1), (J-2), \ldots, -J]$$

thus

$$\varepsilon_{J,v} = \varepsilon_{\text{vib.}} + \varepsilon_{\text{rot.}} = (v+\tfrac{1}{2})\bar{\omega}_e - (v+\tfrac{1}{2})^2 x_e\bar{\omega}_e$$
$$+ BJ(J+1) + (A-B)K^2 \text{ cm}^{-1}. \quad (3.29)$$

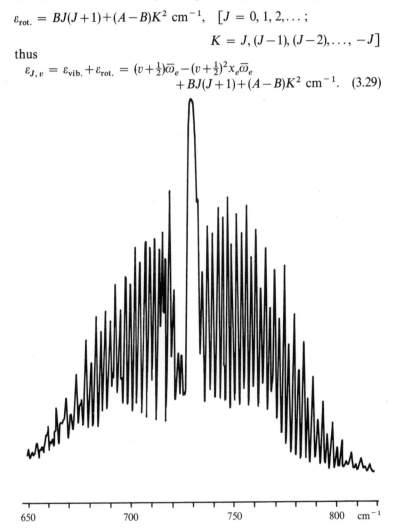

Fig. 3.16: The spectrum of a bending mode of acetylene, HC≡CH, showing
the strong, weak, strong, weak, . . . intensity alternation in the rotational fine
structure due to the nuclear spin of the hydrogen atoms.

This equation assumes, of course, that centrifugal distortion is negligible.

Again it is necessary to divide the vibrations into those which change the dipole (*i*) parallel and (*ii*) perpendicular to the main symmetry axis—which is nearly always the axis about which the "top" rotates. The rotational selection rules differ for the two types.

(*i*) *Parallel Vibrations.* Here the selection rule is:

$$\Delta v = \pm 1, \qquad \Delta J = 0, \pm 1, \qquad \Delta K = 0. \qquad (3.30)$$

Since here $\Delta K = 0$, terms in K will be identical in the upper and lower state and so the spectral frequencies will be independent of K. Thus the situation will be identical to that discussed for the *perpendicular* vibrations of a linear molecule. The spectrum will contain P, Q and R branches with a P, R line spacing of $2B$ (which is unlikely to be resolved) and a strong central Q branch. Such a spectrum, a $\parallel$ band of methyl iodide, CH_3I, is shown in Fig. 3.17.

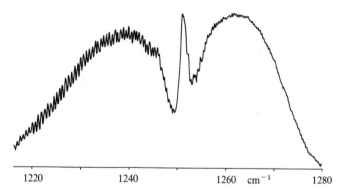

Fig. 3.17: *The parallel stretching vibration, centred at 1251 cm^{-1}, of the symmetric top molecule methyl iodide, CH_3I, showing the typical PQR contour.*

The intensity of the Q branch (relative to lines in the P and R branches) varies with the ratio I_A/I_B; in the limit, when $I_A \to 0$, the symmetric top becomes a linear molecule and the Q branch has zero intensity, as discussed earlier.

(*ii*) *Perpendicular Vibrations.* For these the selection rule is:

$$\Delta v = \pm 1, \qquad \Delta J = 0, \pm 1, \qquad \Delta K = \pm 1. \qquad (3.31)$$

Each of the following expressions are readily derivable for the spectral lines, taking the energy levels of equation (3.29).

(a) $\Delta J = +1$, $\Delta K = \pm 1$ (R branch lines):

$$\Delta\varepsilon = \bar{v}_{spect.} = \bar{\omega}_o + 2B(J+1) + (A-B)(1\pm 2K) \text{ cm}^{-1} \quad (3.32a)$$

(b) $\Delta J = -1$, $\Delta K = \pm 1$ (P branch lines):

$$\bar{v}_{spect.} = \bar{\omega}_o - 2B(J+1) + (A-B)(1\pm 2K) \text{ cm}^{-1} \quad (3.32b)$$

(c) $\Delta J = 0$, $\Delta K = \pm 1$ (Q branch lines):

$$\bar{v}_{spect.} = \bar{\omega}_o + (A-B)(1\pm 2K) \text{ cm}^{-1}. \quad (3.32c)$$

We see then, that this type of vibration gives rise to many sets of P and R branch lines since for each J value there are many allowed values of K ($K = J, J-1, \ldots, -J$). The wings of the spectrum will thus be quite complicated and will not normally be resolvable into separate lines. The Q branch is also complex, since it too will consist of a series of lines on both sides of $\bar{\omega}_o$ separated by $2(A-B)$. This latter term may not be small (and is equal to zero only for *spherical top* molecules which have all their moments of inertia equal). For $A \gg B$ (e.g. CH_3I) the Q branch lines will be well separated and will appear as a series of maxima above the P, R envelope. This spectrum is shown in Fig. 3.18.

It will be noted in this figure that the lines have a distinct periodical variation in intensity—strong, weak, weak, strong, weak, weak, ... This behaviour reminds us of CO_2 and C_2H_2, discussed earlier, in which the presence or absence of nuclear spin altered the relative populations of the rotational levels. In that case, where the molecule had a 2-fold axis of symmetry, the periodicity also was two —strong, weak, strong, weak, ... It is not surprising, therefore, that the 3-fold periodicity, strong, weak, weak, strong, ... seen in CH_3I, arises because of its 3-fold axis of symmetry to rotations about the C—I axis. The appearance of such a spectrum confirms immediately that we are dealing with a molecule containing an XY_3 grouping.

Other Polyatomic Molecules. We shall not go further with the discussion of their detailed spectra here—it suffices to state that the complexity increases, naturally, with the molecular complexity. An excellent treatment is to be found in Herzberg's book, but the subject is not for the beginner in spectroscopy.

Summary. We have seen that the infra-red spectrum of even a

simple diatomic molecule may contain a great many lines, while that
of a polyatom may be extraordinarily complex, even though some
of the details of fine structure are blurred by insufficient resolving

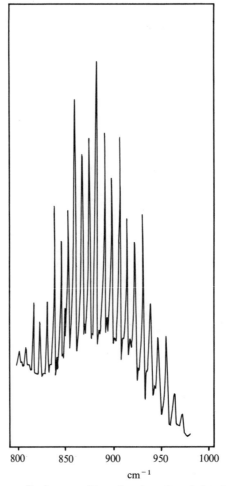

$$cm^{-1}$$

*Fig. 3.18: A perpendicular stretching vibration of methyl iodide showing the
typical Q branch sequence.*

power.   Although in favourable cases much information may be
obtained about bond lengths and angles or at least the general
shape of a molecule, in others even the assignment of observed bands

to particular molecular vibrations is not trivial. Assignments are based mainly on experience with related molecules, on the band contour (from which the type of vibration ($\parallel$ or $\perp$) can usually be deduced), and on the use of Raman spectra (see Chapter 4). Consideration of the symmetry of the molecule is also important because this determines which vibrations are likely to be infra-red active.

Fortunately the usefulness of infra-red spectroscopy extends far beyond the measurement of precise vibrational frequencies and molecular structural features. In the next section we discuss briefly the application of infra-red techniques to chemical analysis —a branch of the subject where it is by no means essential always to be able to assign observed bands precisely.

7. Analysis by Infra-Red Techniques

Because of the $3N-6$ and $3N-5$ rules it is evident that a complex molecule is likely to have an infra-red spectrum exhibiting a large number of normal vibrations. Each normal mode involves some displacement of all, or nearly all, the atoms in the molecule, but while in some of the modes all atoms may undergo approximately the same displacement, in others the displacements of a small group of atoms may be much more vigorous than those of the remainder. Thus we may divide the normal modes into two classes: the *skeletal vibrations*, which involve all the atoms to much the same extent, and the *characteristic group vibrations*, which involve only a small portion of the molecule, the remainder being more or less stationary. We deal with these classes separately.

Skeletal frequencies usually fall in the range 1400–700 cm^{-1} and arise from linear or branched chain structures in the molecule. Thus such groups as

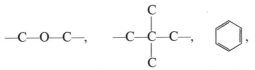

etc., each give rise to several skeletal modes of vibration and hence several absorption bands in the infra-red. It is seldom possible to assign particular bands to specific vibrational modes, but the whole complex of bands observed is highly typical of the molecular structure under examination. Further, changing a substituent (on the chain, or in the ring) usually results in a marked change in the

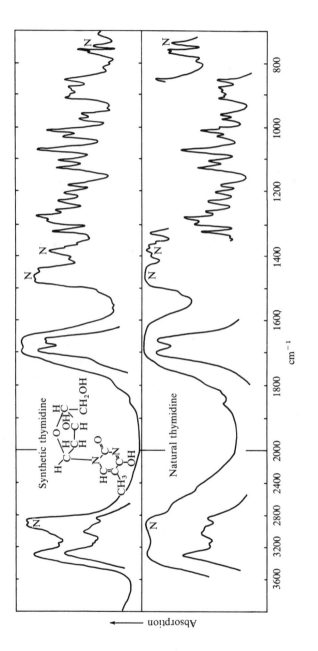

Fig. 3.19(a): *Comparison of the infra-red spectra of natural and synthetic thymidine, to show the use of the fingerprint region.* N=absorption from liquid paraffin (nujol) in which the solid thymidine is suspended. (These spectra and that of Fig. 3.19(b) are reproduced by kind permission of Professor N. Sheppard of the University of East Anglia, Norwich.)

pattern of the absorption bands. Thus these bands are often referred to as the "fingerprint" bands, because a molecule or structural moiety may often be recognized merely from the appearance of this part of the spectrum. An excellent example of this is

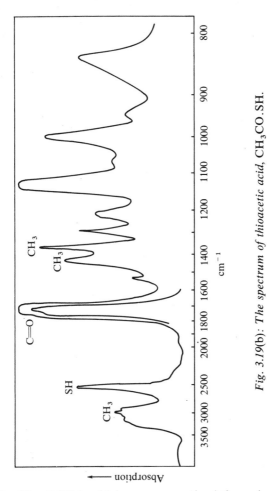

Fig. 3.19(b): The spectrum of thioacetic acid, $CH_3CO.SH$.

shown in Fig. 3.19(a) which compares the infra-red spectra of natural and synthetic thymidine. The remarkably exact correlation between the spectra shows that the synthetic product cannot differ in the slightest degree from the natural substance.

Group frequencies, on the other hand, are usually almost

INFRA-RED SPECTROSCOPY

independent of the structure of the molecule as a whole and, with a few exceptions, fall in the regions well above and well below that of the skeletal modes; Table 3.4 collects some of the data, of which a much more complete selection is to be found in the books by Bellamy and Nakanishi mentioned in the bibliography at the end of

TABLE 3.4: *Characteristic Stretching Frequencies of some Molecular Groups*

Group	Approximate Frequency (cm^{-1})	Group	Approximate Frequency (cm^{-1})
—OH	3600	$>$C=O	1750–1600
—NH$_2$	3400	$>$C=C$<$	1650
≡CH	3300	$>$C=N$\diagdown$	1600
		$>$C—C$<$	
(benzene ring with H)	3060	$>$C—N$<$	1200–1000
		$>$C—O$\diagdown$	
=CH$_2$	3030	$>$C=S	1100
—CH$_3$	2970 (asym. stretch)		
	2870 (sym. stretch)	$>$C—F	1050
	1460 (asym. deform.)		
	1375 (sym. deform.)	$>$C—Cl	725
—CH$_2$—	2930 (asym. stretch)	$>$C—Br	650
	2860 (sym. stretch)		
	1470 (deformation)	$>$C—I	550
—SH	2580		
—C≡N	2250		
—C≡C—	2220		

this chapter. We see that the vibrations of light atoms in terminal groups (e.g. —CH$_3$, —OH, —C≡N, $>$C=O, etc.) are of high frequency, while those of heavy atoms (—C—Cl, —C—Br, metal–metal, etc.) are low in frequency. Their frequencies, and consequently their spectra, are highly characteristic of the group, and can be used for analysis. For example, the —CH$_3$ group gives rise to a symmetric C—H stretching absorption invariably falling between 2850 and 2890 cm^{-1}, an asymmetric stretching frequency at 2940–2980 cm^{-1}, a symmetric deformation (i.e. the opening and closing of the "umbrella") at about 1375 cm^{-1}, and an asymmetric deformation at about 1470 cm^{-1}. Again, the $>$C$=$O group shows a very sharp and intense absorption between 1600 and 1750 cm^{-1}, depending largely on the other substituents of the group. An example of the application of group frequency data is shown in Fig. 3.19(b); this is the spectrum of thioacetic acid— acetic acid in which one oxygen atom has been replaced by sulphur; the question might be asked: is the molecule $CH_3CO.SH$, or $CH_3CS.OH$? The infra-red spectrum gives a very clear answer. It shows a very sharp absorption at about 1730 cm^{-1}, and one at about 2600 cm^{-1}, and these are consistent with the presence of $>$C$=$O and —SH groups, respectively (cf. Table 3.4). Also there is little or no absorption at 1100 cm^{-1} (apart from the general back- ground caused by the skeletal vibrations), thus indicating the ab- sence of $>$C$=$S.

The idea of group vibrations also covers the motions of isolated features of a molecule which have frequencies not too near those of the skeletal vibrations. Thus isolated multiple bonds (e.g. $>$C$=$C$<$, or —C$\equiv$C—) have frequencies which are highly characteristic. When, however, two such groups which, in isolation, have com- parable frequencies, occur together in a molecule, resonance occurs and the group frequencies may be shifted considerably from the expected value. Thus the isolated carbonyl in a ketone $\left(\begin{array}{c} R \\ R \end{array}\!\!>\!\!C=O\right)$ and the $>$C$=$C$<$ double bond, have group frequencies of 1715 and 1650 cm^{-1} respectively; however, when the grouping $>$C$=$C—C$=$O occurs, their separate frequencies are shifted to 1675 and about 1600 cm^{-1} respectively and the intensity of the $>$C$=$C$<$ absorption increases to become comparable with that of the in- herently strong $>$C$=$O band (cf. Fermi resonance, p. 86). Closer coupling of the two groups, as in the ketene radical, $>$C$=$C$=$O, gives rise to absorptions at about 2100 and 1100 cm^{-1}, which are very far removed from the "characteristic" frequencies of the separate groups.

Shifts in group frequencies can arise in other ways too, particularly as the result of interactions between different molecules. Thus the —OH stretching frequency of alcohols is very dependent on the degree of hydrogen bonding, which lengthens and weakens the —OH bond, and hence lowers its vibrational frequency. If the hydrogen bond is formed between the —OH and, say, a carbonyl group, then the latter frequency is also lowered, although to a less extent than the —OH, since hydrogen bonding weakens the $>C=O$ linkage also. However, shifts in group frequency position caused by resonance or intermolecular effects are in themselves highly characteristic and very useful for diagnostic purposes.

In a similar way a change of physical state may cause a shift in the frequency of a vibration, particularly if the molecule is rather polar. In general the more condensed phase gives a lower frequency: $v_{gas} > v_{liquid} \approx v_{solution} > v_{solid}$. Thus in the relatively polar molecule HCl there is a shift of some 100 cm^{-1} in passing from vapour to liquid and a further decrease of 20 cm^{-1} on solidification. Non-polar CO_2, on the other hand, shows negligible shifts in its symmetric vibrations (Fig. 3.11(a) and (b)) but a lowering of some 60 cm^{-1} in v_3 on solidification.

Examination of Table 3.4 shows that there are logical trends in group frequencies, since equation (3.4):

$$\bar{\omega} = \frac{1}{2\pi c} \sqrt{\frac{k}{\mu}} \text{ cm}^{-1}$$

is approximately obeyed. Thus we see that increasing the mass of the atom undergoing oscillation within the group (i.e. increasing μ) tends to decrease the frequency—cf. the series CH, CF, CCl, CBr, or the values for $>C=O$ and $>C=S$. Also, increasing the strength of the bond, and hence increasing the force constant k, tends to increase the frequency, e.g. the series —C—X, —C=X, —C≡X, where X is C, N, or (in the first two fragments) O.

We should at this point consider very briefly the intensities of infra-red bands. We have seen that an infra-red spectrum only appears if the vibration produces a change in the permanent electric dipole of the molecule. It is reasonable to suppose, then, that the more polar a bond, the more intense will be the infra-red spectrum arising from vibrations of that bond. This is generally borne out in practice. Thus the intensities of the $>C=O, >C=N—$ and $>C=C<$ bands decrease in that order, as do those of the —OH,

$>$NH and $>$CH bands. For this reason, too, the vibrations of ionic crystal lattices often give rise to very strong absorptions. We shall see in the next chapter that the reverse is true in Raman spectroscopy—there the less polar (and hence usually more *polarizable*) bonds give the most intense spectral lines.

In summary, then, experience coupled with comparison spectra of known compounds enables one to deduce a considerable amount of structural information from an infra-red spectrum. It should perhaps be mentioned that the *complete* interpretation of the spectrum of a complex molecule can be a very difficult or impossible task. One is usually content to assign the strongest bands and to be able to explain some of the weaker ones as overtones or combinations.

8. Techniques and Instrumentation

8.1. We first deal briefly with each component of the spectrometer as it is usually assembled for infra-red work.

(*i*) Source. The source is always some form of filament which is maintained at red- or white-heat by an electric current. Two common sources are the Nernst filament, consisting of a spindle of rare-earth oxides about 1 in. long and 0·1 in. in diameter, and the "globar" filament, a rod of carborundum, somewhat thicker and longer than the Nernst. The Nernst requires to be pre-heated before it will conduct electricity, but once red-heat is reached the temperature is maintained by the current.

(*ii*) Optical Path and Monochromator. The beam is guided and focused by mirrors silvered on their surfaces. Normally a focus is produced at the point where the sample is to be placed. Ordinary lenses and mirrors are not suitable as glass absorbs strongly over most of the frequencies used. Any windows which are essential (e.g. to contain a sample, or to protect the detector) must be made of mineral salts transparent to infra-red radiation (NaCl and KBr are much used) which have been highly polished in order to reduce scattering to a minimum.

Similarly, the monochromator in some instruments is made of a rock salt or potassium bromide prism, which is rotatable to produce the required frequency in a manner similar to that of Fig. 1.11 (except, of course, that the lens is replaced by a condensing mirror). Modern instruments, however, use a rotatable grating instead of a prism, since this gives much better resolving power.

8

(*iii*) Detector. Two main types are in common use, one sensing the heating effect of the radiation, the other depending on photo-conductivity. In both the greater the effect (temperature or con-ductivity rise) at a given frequency, the greater the transmittance (and the less the absorbance) of the sample at that frequency.

An example of the temperature method is to be found in the Golay cell which is pneumatic in operation. The radiation falls on to a very small cell containing air, and temperature changes are measured in terms of pressure changes within the cell which can be recorded directly as "transmittance". Alternative examples of this type of detector are small, sensitive thermocouples or bolo-meters.

The phenomenon of conductivity in substances is thought to arise as a consequence of the movement of loosely held electrons through the lattice; insulators, on the other hand, have no such loosely bound electrons. Semiconductors are essentially midway between these materials, having no loosely bound electrons in the normal state, but having "conduction bands" or raised electron energy levels into which electrons may be readily excited by the absorption of energy from an outside source. Photo-conductors are a particular class of semiconductors in which the energy required comes from incident radiation, and some materials, such as lead sulphide, have been found sufficiently sensitive to infra-red radia-tion (although only above some 4000 cm^{-1}) that they make excel-lent detectors. The conductivity of the material can be measured continuously by a type of Wheatstone bridge network and, when plotted against frequency, this gives directly the transmittance of the sample.

(*iv*) Sample. For reasons just stated, the sample is held between plates of polished mineral salt rather than glass. Pure liquids are studied in thicknesses of about 0·01 mm, while solutions are usually 0·1–10 mm thick, depending on the dilution. Gas samples at pressures of up to 1 atmosphere or greater are usually contained in glass cells either 5 or 10 cm long, closed at their ends with rock salt windows. Special long-path cells, in which the radiation is re-peatedly reflected up and down the cell, may be used for gases at low pressure, perhaps less than 10 cm Hg.

Solid samples are more difficult to examine because the particles reflect and scatter the incident radiation and transmittance is always low. If the solid cannot be dissolved in a suitable solvent, it is best examined by grinding it very finely in paraffin oil (nujol) and thus forming a suspension, or "mull". This can then be held between salt plates in the same way as a pure liquid or solvent. Provided the refractive indices of liquid and solid phase are not very different, scattering will be slight.

Another technique for handling solids is to grind them very finely with potassium bromide. Under very high pressure this material will flow slightly, and the mixture can usually be pressed into a transparent disk. This may then be placed directly in the infra-red beam in a suitable holder. Although superficially attractive the method is not generally recommended because of the difficulty in obtaining really reproducible results.

8.2. Double and Single Beam Operation. Figure 3.20 shows the spectrum of the atmosphere between 4000 and 400 cm^{-1} taken with a path length of some 2 m—this is not abnormally long for the beam paths in a spectrometer. It is evident that, although H_2O and CO_2 occur in air only in small percentages, their absorbance over much of the spectrum is considerable. Not only would this absorbance have to be subtracted from the spectrum of any sample run under comparable conditions, but, since the percentage of water vapour in the atmosphere is variable, such a "background" spectrum as Fig. 3.20 would have to be run afresh for each sample.

If the regions of these absorbances are not to be denied us in spectroscopic studies, some action must be taken either to remove the H_2O and CO_2 from the air, or to remove the effects of their spectra. It is possible to remove these gases either by complete evacuation of the spectrometer, or by sweeping them out with a current of dry nitrogen, or dry CO_2-free air. The first is not easy since a modern spectrometer may have a volume of some 10 cu. ft and there will be a great many places in its container where leaks may occur. Nor is it ever completely effective, since water vapour proves to be remarkably tenacious and weeks of hard evacuation may be necessary before all the water vapour is desorbed from the surfaces inside the spectrometer. For this reason, also, sweeping with a dry inert gas is not very effective. However, these methods do, quite rapidly, reduce the interference considerably.

The *effects* of this interference can be removed much more simply by using an instrument designed for *double beam* operation. In this, the source radiation is divided into two by means of the mirrors M_1 and M_2 (Fig. 3.21). One beam is brought to a focus at the sample space, while the other follows an exactly equivalent path and is referred to as the reference beam.   The two beams are passed alternately through the monochromator and on to the detector by the rotating sector mirror M_3, and thus the detector "sees" the sample beam and reference beam alternately.   Both beams have

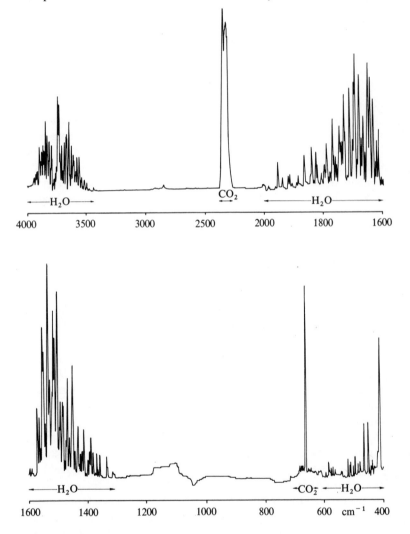

Fig. 3.20: The spectrum of atmospheric water vapour and carbon dioxide.

travelled the same distance through the atmosphere and thus both are reduced in energy to the same extent by absorption by CO_2 and H_2O.

If a sample, capable of absorbing energy from the beam at the particular frequency passed by the monochromator, is now placed in the sample beam, the detector will receive a signal alternating in intensity, since the sample beam carries less energy than the reference beam. It is a simple matter, electronically, to amplify

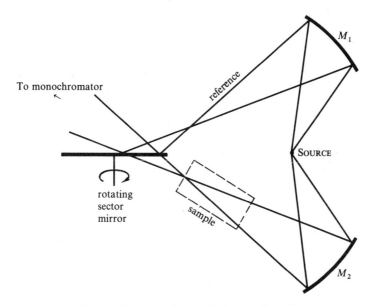

Fig. 3.21: Schematic diagram of a double-beam infra-red spectrometer.

this alternating signal and to arrange that a calibrated attenuator is driven into the reference beam until the signal is reduced to zero, i.e. until sample and reference beams are again balanced. The distance moved by the attenuator is a direct measure of the amount of energy absorbed by the sample.

By balancing sample and reference beams in this way, the absorption of atmospheric CO_2 and H_2O do not appear in the infra-red spectra since both beams are reduced in energy to the same extent. The double beam spectrometer has other advantages, however.

Firstly, it is much simpler to amplify the alternating signal produced than the d.c. signal resulting from a single beam detector.

Further the sector mirror acts in the same way as the Stark modulator for microwave work mentioned in Chapter 2 (Section 5.2) and, by amplifying only that component of the signal having the sector mirror frequency (usually 10–100 c/sec) a great improvement in the signal/noise ratio is obtained.

Thirdly, when examining the spectra of solutions, one can put a cell containing the appropriate quantity of pure solvent into the reference beam, thus eliminating the solvent spectrum from the final trace. On a single beam instrument the solvent spectrum must be taken separately and "subtracted" from the solution spectrum in order to arrive at the spectrum of the substance of interest.

It should be pointed out, however, that a double beam instrument is never *completely* effective in removing traces of water vapour or CO_2 from the spectra. No matter how carefully the instrument is assembled small differences occur in the beam paths and a small residual spectrum results. This can usually be removed, however, by sweeping with dry, inert gas as well as using the double beam principle.

A further, more serious disadvantage which is not always appreciated by users of spectrometers is that the double beam instrument only removes the spectral trace of CO_2 and H_2O; the very strong absorption of energy by these gases still remains in both beams. This means that at some parts of the spectrum the actual amount of energy reaching the detector may be extremely small. Under these conditions, unless the spectrometer is very carefully operated, the spectral trace of a substance may be quite false. Fortunately, regions of very high atmospheric absorption are few and narrow but they should be borne in mind when examining infra-red spectra. This disadvantage can only be removed by sweeping out or evacuating the spectrometer. Similar, but more pronounced, effects occur in regions of strong solvent absorbance when a compensating cell is put in the reference beam.

Bibliography

BARROW, G. M.: *Introduction to Molecular Spectroscopy*, McGraw-Hill Book Co. Inc., 1962

BELLAMY, L. J.: *The Infra-red Spectra of Complex Molecules*, 2nd edn., Methuen, 1958

BERL, W. G. (Ed.): *Physical Methods in Chemical Analysis*, Vol. I, 2nd edn., Academic Press, 1960

HERZBERG, G.: *Molecular Spectra and Molecular Structure;* Vol. 1, *Spectra of Diatomic Molecules*, 2nd edn., Van Nostrand, 1950

HERZBERG, G.: *Molecular Spectra and Molecular Structure;* Vol. 2, *Infra-red and Raman Spectra of Polyatomic Molecules*, Van Nostrand, 1945

KING, G. W.: *Spectroscopy and Molecular Structure*, Holt, Rinehart and Winston, Inc., 1964

MELOAN, C. E.: *Elementary Infra-red Spectroscopy*, Macmillan, 1963

NACHOD, F. C. and W. D. PHILLIPS: *Determination of Organic Structures by Physical Methods*, Vol. 2, Academic Press, 1962

NAKAMOTO, K.: *Infra-red Spectra of Inorganic and Co-ordination Compounds*, John Wiley & Sons, 1963

NAKANISHI, K.: *Infra-red Absorption Spectroscopy*, Holden-Day Inc., 1962

RAO, C. N. R.: *Chemical Applications of Infra-red Spectroscopy*, Academic Press, 1964

WALKER, S. and H. STRAW: *Spectroscopy*, Vol. II, Chapman & Hall, 1962

WIBERLEY, S. E., N. B. COLTHUP and L. H. DALY: *Introduction to Infra-red and Raman Spectroscopy*, Academic Press, 1964

CHAPTER 4

RAMAN SPECTROSCOPY

1. Introduction

When a beam of light is passed through a transparent substance, a small amount of the radiation energy is scattered, the scattering persisting even if all dust particles or other extraneous matter are rigorously excluded from the substance. If monochromatic radiation, or radiation of a very narrow frequency band, is used, the scattered energy will consist almost entirely of radiation of the incident frequency (the so-called *Rayleigh scattering*) but, in addition, certain discrete frequencies above and below that of the incident beam will be scattered; it is this which is referred to as *Raman scattering*.

1.1. Quantum Theory of Raman Effect. The occurrence of Raman scattering may be most easily understood in terms of the quantum theory of radiation. This treats radiation of frequency v as consisting of a stream of particles (called photons) having energy hv where h is Planck's constant. Photons can be imagined to undergo collisions with molecules and, if the collision is perfectly elastic, they will be deflected unchanged. A detector placed to collect energy at right angles to an incident beam will thus receive photons of energy hv, i.e. radiation of frequency v.

However, it may happen that energy is exchanged between photon and molecule during the collision: such collisions are called "inelastic". The molecule can gain or lose amounts of energy only in accordance with the quantal laws, i.e. its energy change, ΔE ergs, must be the difference in energy between two of its allowed states. That is to say, ΔE must represent a change in the vibrational and/or rotational energy of the molecule. If the molecule *gains* energy ΔE ergs, the photon will be scattered with energy $hv - \Delta E$ and the equivalent radiation will have a frequency $v - \Delta E/h$. Conversely if the molecule *loses* energy ΔE, the scattered frequency will be $v + \Delta E/h$.

Radiation scattered with a frequency lower than that of the

110

incident beam is referred to as Stokes' radiation, while that at higher frequency is called anti-Stokes'. Since the former is accompanied by an *increase* in molecular energy (which can always occur, subject to certain selection rules) while the latter involves a *decrease* (which can only occur if the molecule is originally in an excited vibrational or rotational state), Stokes' radiation is generally more intense than anti-Stokes'. Overall, however, the total radiation scattered at any but the incident frequency is extremely small and sensitive apparatus is needed for its study.

1.2. Classical Theory of the Raman Effect; Molecular Polarizability. The classical theory of the Raman effect, while not wholly adequate, is worth consideration since it leads to an understanding of a concept basic to this form of spectroscopy—the polarizability of a molecule. When a molecule is put into a static electric field it suffers some distortion, the positively charged nuclei being attracted towards the negative pole of the field, the electrons to the positive pole. This separation of charge centres causes an *induced electric dipole moment* to be set up in the molecule and the molecule is said to be *polarized*. The size of the induced dipole, μ, depends both on the magnitude of the applied field, E, and on the ease with which the molecule can be distorted. We may write

$$\mu = \alpha E \qquad (4.1)$$

where α is the *polarizability* of the molecule.

Consider first a diatomic molecule, such as H_2, shown at Fig. 4.1(*a*). The polarizability is *anisotropic*, i.e. the electrons forming the bond are more easily displaced by an electric field applied along the bond axis than one across this direction: this may be confirmed experimentally, for example by a study of the absolute intensity of lines in the Raman spectrum, when it is found that the induced dipole moment for a given field applied along the axis is approximately twice as large as that induced by the same field applied across the axis; fields in other directions induce intermediate dipole moments. We can represent the polarizability in various directions most conveniently by drawing a *polarizability ellipsoid*, as in Fig. 4.1(*b*), where the ellipsoid is a three-dimensional surface whose distance from the electrical centre of the molecule (here the centre of gravity also) is proportional to $1/\sqrt{\alpha_i}$, where α_i is the polarizability along the line joining point i on the ellipsoid with the electrical centre. Thus where the polarizability is greatest the axis of the ellipsoid is least, and vice versa. (This representation is chosen because of an analogy with the momentum of a body—the momental ellipsoid is defined similarly using $1/\sqrt{I_i}$, where I_i is the moment of inertia about an axis i.)

Since the polarizability of a diatomic molecule is the same for all directions at right angles to the bond axis, the ellipsoid has a circular cross-section in this direction; thus it is shaped rather like a tangerine.   All diatomic molecules, e.g. CO, HCl, and linear polyatomic molecules, e.g. CO_2, HC≡CH, etc., have ellipsoids of the

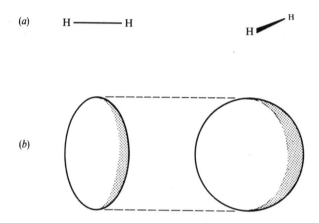

Fig. 4.1: *The hydrogen molecule and its polarizability ellipsoid seen from two directions at right angles.*

same general shape, differing only in the relative sizes of their major and minor axes.

When a sample of such molecules is subjected to a beam of radiation of frequency v the electric field experienced by each molecule varies according to the equation (cf. Chapter 1, equation (1.1)):

$$E = E_0 \sin 2\pi v t \qquad (4.2)$$

and thus the induced dipole also undergoes oscillations of frequency v:

$$\mu = \alpha E = \alpha E_0 \sin 2\pi v t. \qquad (4.3)$$

Such an oscillating dipole emits radiation of its own oscillation frequency and we have immediately in equation (4.3) the classical explanation of Rayleigh scattering.

If, in addition, the molecule undergoes some internal motion, such as vibration or rotation, which *changes the polarizability* periodically, then the oscillating dipole will have superimposed upon it the vibrational or rotational oscillation. Consider, for example, a vibration of frequency $\nu_{vib.}$ which changes the polarizability: we can write

$$\alpha = \alpha_0 + \beta \sin 2\pi\nu_{vib.}t \qquad (4.4)$$

where α_0 is the equilibrium polarizability and β represents the rate of change of polarizability with the vibration. Then we have:

$$\mu = \alpha E = (\alpha_0 + \beta \sin 2\pi\nu_{vib.}t)E_0 \sin 2\pi\nu t$$

or, expanding and using the trigonometric relation,

$$\sin A \sin B = \tfrac{1}{2}\{\cos (A-B) - \cos (A+B)\}$$

we have

$$\mu = \alpha_0 E_0 \sin 2\pi\nu t + \tfrac{1}{2}\beta E_0\{\cos 2\pi(\nu - \nu_{vib.})t - \cos 2\pi(\nu + \nu_{vib.})t\}$$

$$(4.5)$$

and thus the oscillating dipole has frequency components $\nu \pm \nu_{vib.}$ as well as the exciting frequency ν.

It should be carefully noted, however, that if the vibration does not alter the polarizability of the molecule (and we shall later give examples of such vibrations) then $\beta = 0$ and the dipole oscillates only at the frequency of the incident radiation; the same is true of a rotation. Thus we have the general rule:

In order to be Raman active a molecular rotation or vibration must cause some change in a component of the molecular polarizability. A change in polarizability is, of course, reflected by a change in either the *magnitude* or the *direction* of the polarizability ellipsoid.

(This rule should be contrasted with that for infra-red and microwave activity, which is that the molecular motion must produce a change in the electric dipole of the molecule.)

Let us now consider briefly the shapes of the polarizability ellipsoids of more complicated molecules, taking first the bent triatomic molecule H_2O shown at Fig. 4.2(a).   By analogy with the discussion for H_2 given above, we might expect the polarizability surface to be composed of *two* similar ellipsoids, one for each bond. While this may be correct in minute detail, we must remember that the oscillating electric field which we wish to apply for Raman spectroscopy is usually that of radiation in the visible or ultra-violet region, i.e. having a wave-length of some 10,000–500 Å (cf. Fig. 1.4); molecular bonds, on the other hand, have dimensions of only one or two Å, so we cannot expect our radiation to probe the

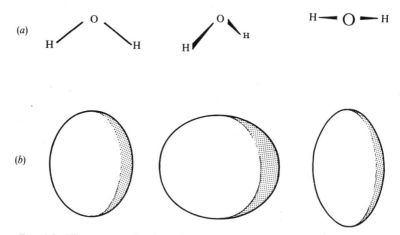

Fig. 4.2: *The water molecule and its polarizability ellipsoid seen along the three co-ordinate axes.*

finer details of bond polarizability—even the hardest of X-rays can scarcely do that.   Instead the radiation can only sense the average polarizability in various directions through the molecule, and the polarizability ellipsoid, it may be shown, is always a true ellipsoid— i.e. a surface having *all* sections elliptical (or possibly circular).   In the particular case of H_2O the polarizability is found to be different along all three of the major axes of the molecule (which lie along the line in the molecular plane bisecting the HOH angle, at right angles to this in the plane, and perpendicularly to the plane), and so all three of the ellipsoidal axes are also different; the ellipsoid is sketched in various orientations in Fig. 4.2(b).   Other such molecules, e.g. H_2S or SO_2, have similarly shaped ellipsoids but with different dimensions.

Symmetric top molecules, because of their axial symmetry, have polarizability ellipsoids rather similar to those of linear molecules, i.e. with a circular cross-section at right angles to their axis of symmetry. It should be stressed, however, that sections in other planes are truly *elliptical*. For a molecule such as chloroform, $CHCl_3$ (Fig. 4.3(a)), where the chlorine atoms are bulky, the usual tendency is to draw the polarizability surface as egg-shaped, fatter at the chlorine-containing end. This is not correct; the polarizability ellipsoid for chloroform is shown at Fig. 4.3(b) where it will be seen that, since the polarizability is greater across the symmetry

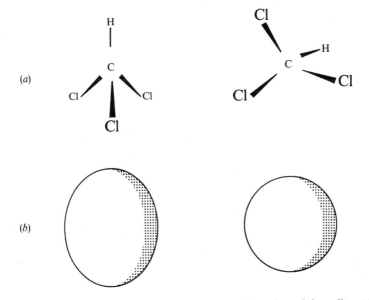

Fig. 4.3: *The chloroform molecule,* $CHCl_3$, *and its polarizability ellipsoid from across and along the symmetry axis.*

axis, the *minor* axis of the ellipsoid lies in this direction. Similar molecules are, e.g. CH_3Cl and NH_3, etc. (although the latter fortuitously has a virtually spherical "ellipsoid").

Finally, spherical top molecules, such as CH_4, CCl_4, SiH_4, etc., have spherical polarizability surfaces, since they are completely isotropic as far as incident radiation is concerned.

We are now in a position to discuss in detail the Raman spectra of various types of molecules. Since we shall be dealing with rotational and vibrational changes it is evident that expressions for the energy levels and for many of the allowed transitions will be identical with those already discussed in the previous two chapters. For clarity we shall repeat any such expressions but not rederive them, being content to give a cross-reference to where their derivation may be found.

2. Pure Rotational Raman Spectra

2.1. Linear Molecules. The rotational energy levels of linear molecules have already been stated (cf. equation (2.24)):

$$\varepsilon_J = BJ(J+1) - DJ^2(J+1)^2 \text{ cm}^{-1} \quad (J = 0, 1, 2, \ldots)$$

but, in Raman spectroscopy, the precision of the measurements does not normally warrant the retention of the term involving D, the centrifugal distortion constant. Thus we take the simpler expression:

$$\varepsilon_J = BJ(J+1) \text{ cm}^{-1} \quad (J = 0, 1, 2, \ldots) \tag{4.6}$$

to represent the energy levels.

Transitions between these levels follow the formal selection rule:

$$\Delta J = 0, \text{ or } \pm 2 \text{ only} \tag{4.7}$$

which is to be contrasted with the corresponding selection rule for microwave spectroscopy, $\Delta J = \pm 1$, given in equation (2.17). The fact that in Raman work the rotational quantum number changes by two units rather than one is connected with the symmetry of the polarizability ellipsoid. For a linear molecule, such as is depicted in Fig. 4.1, it is evident that during end-over-end rotation the ellipsoid presents the same appearance to an observer *twice* in every complete rotation. It is equally clear that rotation about the bond axis produces no change in polarizability, hence, as in infra-red and microwave spectroscopy, we need concern ourselves only with end-over-end rotations.

If, following the usual practice, we define ΔJ as $(J_{\text{upper state}} - J_{\text{lower state}})$ then we can ignore the selection rule $\Delta J = -2$ since, for a pure rotational change, the upper state quantum number must necessarily be greater than that in the lower state. Further the "transition" $\Delta J = 0$ is trivial since this represents no change in the molecular energy and hence Rayleigh scattering only.

Combining, then, $\Delta J = +2$ with the energy levels of equation (4.6) we have:

$$\Delta \varepsilon = \varepsilon_{J'=J+2} - \varepsilon_{J''=J}$$

$$= B(4J+6) \text{ cm}^{-1}. \qquad (4.8)$$

Since $\Delta J = +2$, we may label these lines S branch lines (cf. Chapter 3, Section 2) and write

$$\Delta \varepsilon_S = B(4J+6) \text{ cm}^{-1} \quad (J = 0, 1, 2, \ldots), \qquad (4.9)$$

where J is the rotational quantum number in the *lower* state.

Thus if the molecule gains rotational energy from the photon during collision we have a series of S branch lines to the low wavenumber side of the exciting line (Stokes' lines), while if the molecule loses energy to the photon the S branch lines appear on the high wavenumber side (anti-Stokes' lines). The wavenumbers of the corresponding spectral lines are given by:

$$\bar{\nu}_S = \bar{\nu}_{\text{ex.}} \pm \Delta \varepsilon_S = \bar{\nu}_{\text{ex.}} \pm B(4J+6) \text{ cm}^{-1}, \qquad (4.10)$$

where the plus sign refers to anti-Stokes' lines, the minus to Stokes', and $\bar{\nu}_{\text{ex.}}$ is the wavenumber of the exciting radiation.

The allowed transitions and the Raman spectrum arising are shown schematically in Fig. 4.4. Each transition is labelled according to its lower J value and the relative intensities of the lines are indicated assuming that the population of the various energy levels varies according to equation (2.21) and Fig. 2.7. In particular it should be noted here that Stokes' and anti-Stokes' lines have comparable intensity because many rotational levels are populated and hence downward transitions are approximately as likely as upward ones.

When the value $J=0$ is inserted into equation (4.10) it is seen immediately that the separation of the first line from the exciting line is $6B$ cm^{-1}, while the separation between successive lines is $4B$ cm^{-1}. For diatomic and light triatomic molecules the rotational Raman spectrum will normally be resolved and we can immediately obtain a value of B, and hence the moment of inertia and bond lengths for such molecules. If we recall that homonuclear diatomic molecules (e.g. O_2, H_2) give no infra-red or microwave spectra since they possess no dipole moment, whereas they *do* give a rotational Raman spectra, we see that the Raman technique yields structural data unobtainable from the techniques previously discussed. It is thus complementary to microwave and infra-red studies, not merely confirmatory.

It should be mentioned that, if the molecule has a centre of symmetry (as, for example, do H_2, O_2, CO_2), then the effects of

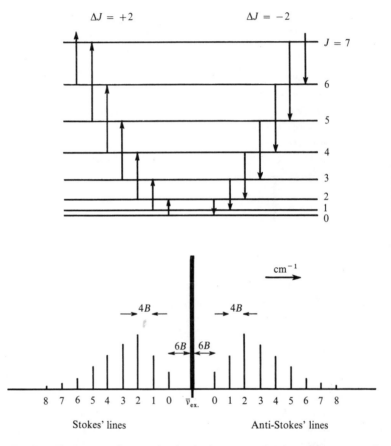

Fig. 4.4: *The rotational energy levels of a diatomic molecule and the rotational Raman spectrum arising from transitions between them. Spectral lines are numbered according to their* lower *J values.*

nuclear spin will be observed in the Raman as in the infra-red. Thus for O_2 and CO_2 (since the spin of oxygen is zero) every alternate rotational level is absent; for example, in the case of O_2, every level with *even* J values is missing, and thus every transition labelled $J = 0, 2, 4, \ldots$ in Fig. 4.4 is also completely missing from the spectrum.   In the case of H_2, and other molecules composed of nuclei with non-zero spin, the spectral lines show an alternation of intensity.

Linear molecules with more than three heavy atoms have large moments of inertia and their rotational fine structure is often unresolved in the Raman spectrum.   Direct structural information is not, therefore, obtainable, but we shall see shortly that, taken in conjunction with the infra-red spectrum, the Raman can still yield much very useful information.

2.2. *Symmetric Top Molecules.*   The polarizability ellipsoid for a typical symmetric top molecule, e.g. $CHCl_3$, was shown in Fig. 4.3(*b*).   Plainly rotation about the top axis produces no change in the polarizability, but end-over-end rotations will produce such a change.

From equation (2.38) we have the energy levels:

$$\varepsilon_{J,K} = BJ(J+1) + (A-B)K^2 \text{ cm}^{-1} \quad (J = 0, 1, 2, \ldots;$$
$$K = \pm J, \pm(J-1), \ldots). \quad (4.11)$$

The selection rules for Raman spectra are:

$$\left.\begin{array}{l} \Delta K = 0 \\[4pt] \Delta J = 0, \pm 1, \pm 2 \quad \text{(except for } K = 0 \text{ states} \\ \qquad\qquad\qquad \text{when } \Delta J = \pm 2 \text{ only).} \end{array}\right\} \quad (4.12)$$

K, it will be remembered, is the rotational quantum number for axial rotation, so the selection rule $\Delta K = 0$ implies that changes in the angular momentum about the top axis will not give rise to a Raman spectrum—such rotations are, as mentioned previously, Raman inactive.   The restriction of ΔJ to ± 2 for $K = 0$ states means effectively that ΔJ cannot be ± 1 for transitions involving the ground state $(J = 0)$ since $K = \pm J, \pm(J-1), \ldots, 0$.   Thus for all J values other than zero, K also may be different from zero and $\Delta J = \pm 1$ transitions are allowed.

Restricting ourselves, as before, to positive ΔJ we have the two cases:

(*i*) $\Delta J = +1$ (*R* branch lines)

$$\Delta\varepsilon_R = \varepsilon_{J'=J+1} - \varepsilon_{J''=J}$$
$$= 2B(J+1) \text{ cm}^{-1} \quad (J = 1, 2, 3, \ldots \text{ (but } J \neq 0)) \quad (4.13a)$$

9

(*ii*) $\Delta J = +2$ (*S* branch lines)

$$\Delta\varepsilon_S = \varepsilon_{J'=J+2} - \varepsilon_{J''=J}$$
$$= B(4J+6)\ \text{cm}^{-1} \quad (J = 0, 1, 2, \ldots). \tag{4.13b}$$

Thus we shall have two series of lines in the Raman spectrum:

$$\left.\begin{aligned}
\bar{v}_R &= \bar{v}_{\text{ex.}} \pm \Delta\varepsilon_R = \bar{v}_{\text{ex.}} \pm 2B(J+1)\ \text{cm}^{-1} \quad (J = 1, 2, \ldots) \\
\bar{v}_S &= \bar{v}_{\text{ex.}} \pm \Delta\varepsilon_S = \bar{v}_{\text{ex.}} \pm B(4J+6)\ \text{cm}^{-1} \quad (J = 0, 1, 2, \ldots)
\end{aligned}\right\} \tag{4.14}$$

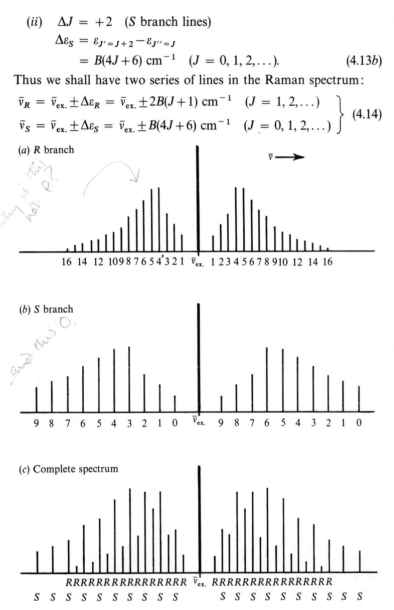

Fig. 4.5: Rotational Raman spectrum of a symmetric top molecule. The R and S branch lines are shown separately in (a) and (b) respectively, with the total spectrum in (c).

These series are sketched separately in Fig. 4.5(a) and (b), where each line is labelled with its corresponding *lower* J value.   In the R branch, lines appear at $4B, 6B, 8B, 10B, \dots$ cm^{-1} from the exciting line, while the S branch series occurs at $6B, 10B, 14B, \dots$ cm^{-1}. The complete spectrum, shown at Fig. 4.5(c) illustrates how every alternate R line is overlapped by an S line.   Thus a marked intensity alternation is to be expected which, it should be noted, is not connected with nuclear spin statistics.

2.3. Spherical Top Molecules; Asymmetric Top Molecules. Examples of spherical top molecules are those with tetrahedral symmetry such as methane, CH_4 or silane, SiH_4.   The polarizability ellipsoid for such molecules is a spherical surface and it is evident that rotation of this ellipsoid will produce no change in polarizability.   Therefore the pure rotations of spherical top molecules are completely inactive in the Raman.

Normally *all* rotations of asymmetric top molecules, on the other hand, are Raman active.   Their Raman spectra are thus quite complicated and will not be dealt with in detail here; it suffices to say that, as in the microwave region, the spectra may often be interpreted by considering the molecule as intermediate between the oblate and prolate types of symmetric top.

3. Vibrational Raman Spectra

3.1. Raman Activity of Vibrations.   If a molecule has little or no symmetry it is a very straightforward matter to decide whether its vibrational modes will be Raman active or inactive: in fact, it is usually correct to assume that *all* its modes are Raman active. But when the molecule has considerable symmetry it is not always easy to make the decision, since it is sometimes not clear, without detailed consideration, whether or not the polarizability changes during the vibration.

We consider first the simple asymmetric top molecule H_2O whose polarizability ellipsoid was shown in Fig. 4.2.   In Fig. 4.6 we illustrate at (a), (b) and (c) respectively the three fundamental modes v_1, v_2 and v_3, sketching for each mode the equilibrium configuration in the centre with the extreme positions to right and left.   The approximate shapes of the corresponding polarizability ellipsoids are also shown.

During the symmetric stretch, in (a), the molecule as a whole

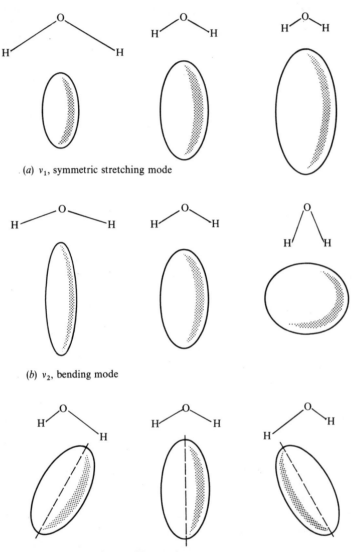

(a) v_1, symmetric stretching mode

(b) v_2, bending mode

(c) v_3, asymmetric stretching mode

Fig. 4.6: The change in size, shape or direction of the polarizability ellipsoid of the water molecule during each of its three vibrational modes. The centre column shows the equilibrium position of the molecule while to right and left are the extremes of each vibration.

increases and decreases in size; now when a bond is stretched, the electrons forming it are less firmly held by the nuclei and so the bond becomes more polarizable. Thus the polarizability ellipsoid of H_2O may be expected to decrease in size while the bonds stretch, and to increase while they compress, but to maintain an approximately constant shape. On the other hand, while undergoing the bending motion, in (b), it is the shape of the ellipsoid which changes most; thus if we imagine vibrations of very large amplitude, at one extreme (on the left) the molecule approaches the linear configuration with a horizontal axis, while at the other extreme (on the right) it approximates to a diatomic molecule (if the two H atoms are almost coincidental), with a vertical axis. Finally in (c) we have the asymmetric stretching motion, v_3, where both the size and shape remain approximately constant, but the direction of the major axis (shown dotted) changes markedly. Thus all three vibrations involve obvious changes in at least one aspect of the polarizability ellipsoid, and all are Raman *active*.

Now consider the linear triatomic molecule CO_2, whose three fundamental vibrational modes have been shown in Fig. 3.11; in Fig. 4.7 we illustrate the extreme and equilibrium configurations of the molecule and their approximate polarizability ellipsoids. The question of the Raman behaviour of the symmetric stretching mode, v_1, is easily decided—during the motion the molecule changes size, and so there is a corresponding fluctuation in the size of the ellipsoid; the motion is thus Raman *active*. It might be thought that the v_2 and v_3 vibrations are also Raman active, because the molecule changes shape during each vibration and hence, presumably, so does the ellipsoid; however, both these modes are observed to be Raman *inactive*. We must, then, consider this example rather more carefully.

To do this it is usual to discuss the change of polarizability with some *displacement co-ordinate*, normally given the symbol ξ; thus for a stretching motion, ξ is a measure of the extension (positive ξ) or compression (negative ξ) of the bond under consideration; while for a bending mode, ξ measures the displacement of the bond angle from its equilibrium value, positive and negative ξ referring to opposite displacement directions.

Consider, as an example, the v_1 stretch of carbon dioxide sketched in Fig. 4.7(a). If the equilibrium value of the polarizability is α_0 (centre picture) then when the bonds stretch (ξ positive), α decreases

124 RAMAN SPECTROSCOPY

(remember that the extent of the ellipsoid measures the *reciprocal* of α), while when the bonds contract (negative ζ) α increases. Thus we can sketch the variation of α with ζ as in Fig. 4.8(*a*). The details of the curve are not important since we are concerned only with *small* displacements; it is plain that near the equilibrium position

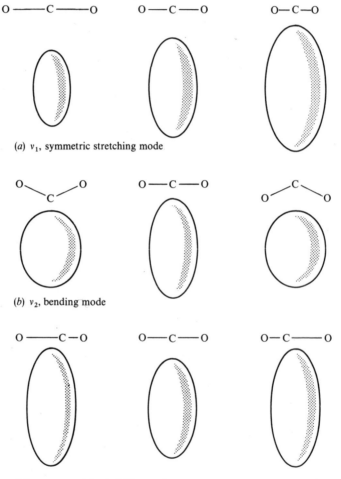

(*a*) v_1, symmetric stretching mode

(*b*) v_2, bending mode

(*c*) v_3, asymmetric stretching mode

Fig. 4.7: The shape of the polarizability ellipsoid of the carbon dioxide molecule during its vibrations.

($\xi = 0$) the curve has a distinct slope, i.e. $d\alpha/d\xi \neq 0$ at $\xi = 0$. Thus for small displacements the motion produces a change in polarizability and is therefore Raman *active*.

If we now consider the situation for v_2, the bending motion of Fig. 4.7(b), we can count a downwards displacement of the oxygen atoms as negative ξ and an upwards displacement as positive. Although it is not clear from the diagrams whether the motion causes an increase or a decrease in polarizability (actually it is an *increase*) it *is* plain that the change is exactly the same for both positive and negative ξ. Thus we can plot α against ξ as in Fig. 4.8(b) with, as before, $\alpha = \alpha_0$ at $\xi = 0$. Now for small displacements

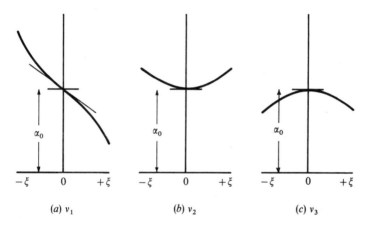

(a) v_1 (b) v_2 (c) v_3

Fig. 4.8: *The variation of the polarizability, α, with the displacement co-ordinate, ξ, during the three vibrational modes of the carbon dioxide molecule.*

we evidently have $d\alpha/d\xi = 0$ and hence for *small displacements* there is effectively no change in the polarizability and the motion is Raman *inactive*.

Exactly the same argument applies to the asymmetric stretch, v_3, shown at Fig. 4.7(c). Here the polarizability decreases equally for positive and negative ξ, so the plot of polarizability against ξ has the appearance of Fig. 4.8(c). Again $d\alpha/d\xi = 0$ for small displacements and the motion is Raman *inactive*.

We could have followed the same reasoning for the three vibrations of water discussed previously. In each case we would have discovered that the α vs. ξ curve has the general shape of Fig. 4.8(a) or its mirror image; in other words, in each case $d\alpha/d\xi \neq 0$ and the motion is Raman active. In general, however, the slopes of the three curves would be different at $\xi = 0$, i.e. $d\alpha/d\xi$ would have different values. Since we have seen that the Raman spectrum is forbidden for $d\alpha/d\xi = 0$ but allowed for $d\alpha/d\xi \neq 0$ we can imagine that the "degree of allowedness" varies with $d\alpha/d\xi$. Thus if the polarizability curve has a large slope at $\xi = 0$ the Raman line will be strong, if the slope is small it will be weak, and if zero, not allowed at all. From this stems the following very useful general rule:

symmetric vibrations give rise to intense Raman lines; non-symmetric ones are usually weak and sometimes unobservable.

In particular, a bending motion usually yields only a very weak Raman line; for example the v_2 motion of H_2O (Fig. 3.6(b)), although allowed in the Raman, has not been observed, nor has v_3, for which $d\alpha/d\xi$ is also small.

3.2. Rule of Mutual Exclusion. A further extremely important general rule has been established whose operation may be exemplified by carbon dioxide. We can summarize our conclusions about the Raman and infra-red activities of the fundamental vibrations of this molecule in Table 4.1:

TABLE 4.1: Raman and Infra-Red Activities of Carbon Dioxide

Mode of Vibration of CO_2	Raman	Infra-Red
v_1: symmetric stretch	active	inactive
v_2: bend	inactive	active
v_3: asymmetric stretch	inactive	active

and we see that, for this molecule, no vibration is simultaneously active in both Raman and infra-red. The corresponding general rule is:

Rule of mutual exclusion: if a molecule has a *centre of symmetry* then Raman active vibrations are infra-red inactive, and vice-versa. If there is no centre of symmetry then some (but not necessarily all) vibrations may be both Raman and infra-red active.

The converse of this rule is also true, i.e. the observance of Raman and infra-red spectra showing no common lines implies that the molecule has a centre of symmetry; but here caution is necessary since, as we have already seen, a vibration may be Raman active but too weak to be observed. However, if some vibrations are observed to give coincident Raman and infra-red absorptions it is certain that the molecule has no centre of symmetry. Thus extremely valuable structural information is obtainable by comparison of the Raman and infra-red spectra of a substance; we shall show some examples of this in Section 5.

3.3. Overtone and Combination Vibrations. Without detailed consideration of the symmetry of a molecule and of its various modes of vibration, it is no easy matter to predict the activity, either in Raman or infra-red, of its overtone and combination modes. The nature of the problem can be seen by considering v_1 and v_2 of carbon dioxide; the former is Raman active only, the latter infra-red active. What, then, of the activity of $v_1 + v_2$? In fact it is only infra-red active, but this is not at all obvious merely from considering the dipole or polarizability changes during the motions. Again, when discussing Fermi resonance (Chapter 3, Section 5.2) we chose as an example the resonance of v_1 and $2v_2$ of carbon dioxide in the *Raman* effect. Thus $2v_2$ is Raman active although the fundamental v_2 is only infra-red active.

We shall not attempt here to discuss this matter further, being content to leave the reader with a warning that the activity or inactivity of a fundamental in a particular type of spectroscopy does not necessarily imply corresponding behaviour of its overtones or combinations, particularly if the molecule has considerable symmetry. A more detailed discussion is to be found in Herzberg's book *Infra-red and Raman Spectra* and others mentioned in the bibliography.

3.4. Vibrational Raman Spectra. The structure of vibrational Raman spectra is easily discussed. For every vibrational mode we can write an expression of the form:

$$\varepsilon = \bar{\omega}_e(v+\tfrac{1}{2}) - \bar{\omega}_e x_e(v+\tfrac{1}{2})^2 \text{ cm}^{-1} \quad (v = 0, 1, 2, \ldots) \quad (4.15)$$

where, as before (cf. equation (3.12)), $\bar{\omega}_e$ is the equilibrium vibrational frequency expressed in wavenumbers and x_e is the anharmonicity constant. Such an expression is perfectly general, whatever the shape of the molecule or the nature of the vibration. Quite general, too, is the selection rule:

$$\Delta v = 0, \pm 1, \pm 2, \ldots \quad (4.16)$$

which is the same for Raman as for infra-red spectroscopy, the probability of $\Delta v = \pm 2, \pm 3, \ldots$ decreasing rapidly.

Particularizing, now, to Raman active modes, we can apply the selection rule (4.16) to the energy level expression (4.15) and obtain the transition energies (cf. equations (3.15)):

$$\left.\begin{array}{l} v = 0 \rightarrow v = 1 : \Delta\varepsilon_{\text{fundamental}} = \bar{\omega}_e(1-2x_e) \text{ cm}^{-1} \\[2mm] v = 0 \rightarrow v = 2 : \Delta\varepsilon_{\text{overtone}} \quad = 2\bar{\omega}_e(1-3x_e) \text{ cm}^{-1} \\[2mm] v = 1 \rightarrow v = 2 : \Delta\varepsilon_{\text{hot}} \quad\quad = \bar{\omega}_e(1-4x_e) \text{ cm}^{-1}, \text{ etc.} \end{array}\right\} \quad (4.17)$$

Since the Raman scattered light is, in any case, of low intensity we can ignore completely all the weaker effects such as overtones and "hot" bands, and restrict our discussion merely to the fundamentals. This is not to say that active overtones and hot bands cannot be observed, but they add little to the discussion here.

We would expect Raman lines to appear at distances from the exciting line corresponding to each active fundamental vibration. In other words we can write:

$$\bar{\nu}_{\text{fundamental}} = \bar{\nu}_{\text{ex.}} \pm \Delta\varepsilon_{\text{fundamental}} \text{ cm}^{-1} \quad\quad (4.18)$$

where the minus sign represents the Stokes' lines (i.e. for which the molecule has gained energy at the expense of the radiation) and the plus sign refers to the anti-Stokes' lines. The latter are often too weak to be observed, since as we saw earlier (cf. p. 67) very few of the molecules exist in the $v=1$ state at normal temperatures.

The vibrational Raman spectrum of a molecule is, then, basically simple. It will show a series of reasonably intense lines to the low frequency side of the exciting line with a much weaker, mirror-image series on the high frequency side. The separation of each line from the centre of the exciting line gives immediately the Raman active fundamental vibration frequencies of the molecule.

As an example we illustrate the Raman spectrum of chloroform, $CHCl_3$, a symmetric top molecule (Fig. 4.9(a)). The exciting line in this case is the very intense mercury line at 4358·3 Å, and a wavenumber scale is drawn from this line as zero. Raman lines appear at 262, 366, 668, 761, 1216 and 3019 cm^{-1} on the low frequency (Stokes') side of the exciting line while the line at 262 cm^{-1} on the high frequency (anti-Stokes') side is included for a comparison of its intensity.

For comparison also we show at Fig. 4.9(b) the *infra-red* spectrum

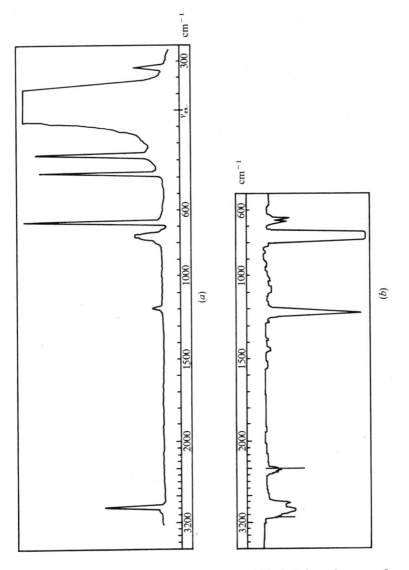

Fig. 4.9: Comparison between (a) the Raman and (b) the infra-red spectra of chloroform, $CHCl_3$, demonstrating the coincidence of bands. In both spectra there is a scale change at 2000 cm^{-1}. The weak absorption at about 2400 cm^{-1} in the infra-red is an overtone of the very strong 1200 cm^{-1} band.   (Thanks are due to Dr. Riley of the Brighton College of Technology for assistance in obtaining the above Raman spectrum.)

of the same molecule. The range of the instrument used pre-
cluded measurements below 600 cm^{-1}, but we see clearly that
strong (and hence fundamental) lines appear in the spectrum at
wavenumbers corresponding very precisely with those of lines in
the Raman spectrum.

For this molecule, containing five atoms, nine fundamental
vibrations (i.e. $3N-6$) are to be expected. The molecule has considerable symmetry, however, and three of these vibrations are
doubly degenerate (see Herzberg: *Infra-red and Raman Spectra* for
details) leaving six different fundamental absorptions; we see that
these are all active in both the infra-red and Raman. The immediate conclusion, not at all surprising, is that the molecule has no
centre of symmetry.

3.5. Rotational Fine Structure. We need not consider in detail
the rotational fine structure of Raman spectra in general, if only
because such fine structure is rarely resolved, except in the case of
diatomic molecules. For the latter we can write the vibration–
rotation energy levels (cf. equation (3.18)) as:

$$\varepsilon_{J,v} = \bar{\omega}_e(v+\tfrac{1}{2})-\bar{\omega}_e x_e(v+\tfrac{1}{2})^2 + BJ(J+1) \text{ cm}^{-1}$$

$$(v = 0, 1, 2,\ldots, J = 0, 1, 2,\ldots) \quad (4.19)$$

where, as before in Raman, we ignore centrifugal distortion. For
diatomic molecules, the selection rule for J is $\Delta J=0, \pm 2$ (Section
2.1) and, combining this with the vibrational change $v=0\rightarrow v=1$,
we have:

$$\Delta J = 0: \quad \Delta\varepsilon_Q = \bar{v}_o \text{ cm}^{-1} \quad \text{(for all } J) \quad (4.20)$$

$$\Delta J = +2: \quad \Delta\varepsilon_S = \bar{v}_o + B(4J+6) \quad (J = 0, 1, 2,\ldots)$$
$$\Delta J = -2: \quad \Delta\varepsilon_O = \bar{v}_o - B(4J+6) \quad (J = 2, 3, 4,\ldots) \quad \Bigg\} \quad (4.21)$$

where we write $\bar{v}_o$ for $\bar{\omega}_e(1-2x_e)$ and use the subscripts O, Q and S
to refer to the O branch lines ($\Delta J=-2$), Q branch lines ($\Delta J=0$) and
S branch lines ($\Delta J=+2$), respectively.

Stokes' lines (i.e. lines to *low* frequency of the exciting radiation)
will occur at wavenumbers given by:

$$\bar{v}_Q = \bar{v}_{\text{ex.}} - \Delta\varepsilon_Q = \bar{v}_{\text{ex.}} - \bar{v}_o \text{ cm}^{-1} \quad \text{(for all } J)$$

$$\bar{v}_O = \bar{v}_{\text{ex.}} - \Delta\varepsilon_O = \bar{v}_{\text{ex.}} - \bar{v}_o + B(4J+6) \text{ cm}^{-1} \quad (J = 2, 3, 4,\ldots)$$

$$\bar{v}_S = \bar{v}_{\text{ex.}} - \Delta\varepsilon_S = \bar{v}_{\text{ex.}} - \bar{v}_o - B(4J+6) \text{ cm}^{-1} \quad (J = 0, 1, 2,\ldots).$$

The spectrum arising is sketched in Fig. 4.10 where, for completeness, the pure rotation lines in the immediate vicinity of the exciting
line are also shown.   The presence of the strong Q branch in the
Raman spectrum is to be noted and compared with the P and R
branches only which occur for a diatomic molecule in the infra-red
(cf. the spectrum of carbon monoxide in Fig. 3.7).   The analysis of
the O and S branches in the Raman spectrum to give a value for

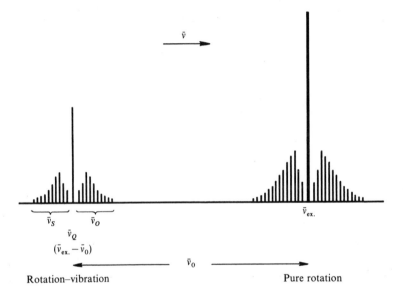

Fig. 4.10: The pure rotation and the rotation–vibration spectrum of a diatomic
molecule having a fundamental vibration frequency of $\bar{\nu}_0$ cm^{-1}.   Stokes'
lines only are shown.

B and hence for the moment of inertia and bond length is straightforward.

Much weaker anti-Stokes' lines will occur at the same distance
from, but to high frequency of, the exciting line.

The resolution of Raman spectra is not sufficient to warrant the
inclusion of finer details such as centrifugal distortion or the break-
down of the Born–Oppenheimer approximation which were discussed in Chapter 3 for the corresponding infra-red spectra.

For larger molecules we can, in fact, ignore the rotational fine
structure altogether since it is not resolved.   Even the O and S (or
O, P, R and S) band contours are seldom observed since they are

very weak compared with the Q branch.   Thus, while the infra-red spectrum of chloroform, shown in Fig. 4.9(b) shows distinct PR or PQR structure on some bands, the Raman spectrum, with the possible exception of the band at 760 cm^{-1}, shows only the strong Q branches.   While some information is denied us in Raman spectra because of this, it does represent a considerable simplification of the overall appearance of such spectra.

In Table 4.2 we collect together some of the information on bond lengths and vibration frequencies which have been obtained from vibrational–rotational Raman spectra.   In the case of CS_2 and CH_4 the ‧symmetrical stretching modes only are given since the wavenumbers of the other modes are determined from infra-red techniques.

TABLE 4.2: Some Molecular Data Determined by
Raman Spectroscopy

Molecule	Bond Length (Å)	Vibration (cm^{-1})
H_2	0.7413 ± 0.0001	4395.2
N_2	1.0976 ± 0.0001	2359.6
F_2	1.418 ± 0.001	802.1
CS_2	1.553 ± 0.005	656.6 (symm. stretch)
CH_4	1.094 ± 0.001	2914.2 (symm. stretch)

4. Polarization of Light and the Raman Effect

4.1. The Nature of Polarized Light.   It is well known that when a beam of light is passed through a Nicol prism or a piece of crystal filter (e.g. polaroid) the only light passing has its electrical (or magnetic) vector confined to a particular plane; it is *plane polarized light*.   Although superficially this light is indistinguishable from ordinary (or unpolarized) light, it has a very important property which can be demonstrated by using a second Nicol prism or crystal filter.   When previously polarized light falls on the second polarizing device (now called the "analyser") it will be passed with undiminished intensity only if the polarizing axes of the two prisms or crystal sheets are parallel to each other.   At any other orientation of these axes the intensity passed will decrease until, when the axes are perpendicular, no light at all passes through the analyser. Thus the analyser serves both to detect polarized light and to determine its plane of polarization.

If the light incident upon the analyser is only partially polarized —i.e. if the majority, but not all, of the rays have their electric vectors parallel to a given plane—then the light will not be *completely* extinguished at any orientation of the analyser; its intensity will merely go through a minimum when the analyser is perpendicular to the plane of maximum polarization. We could, then, measure the degree of polarization in terms of the intensity of light transmitted parallel and perpendicular to this plane; it is more convenient, however, to measure the *degree of depolarization, ρ*, as:

$$\rho = I_\perp/I_\parallel \qquad (4.22)$$

where $I_\parallel$ is the maximum and $I_\perp$ the minimum intensity passed by the analyser. Thus for completely plane polarized light $I_\perp = 0$ and hence the degree of depolarization is zero also; for completely unpolarized (i.e. ordinary) light, $I_\perp = I_\parallel$ and $\rho = 1$. For intermediate degrees of polarization ρ lies between 0 and 1.

The relevance of this to Raman spectroscopy is that lines in some Raman spectra are found to be plane polarized to different extents even though the exciting radiation is completely depolarized. The reason for this is most easily seen if we consider the vibrations of spherical top molecules.

4.2. Vibrations of Spherical Top Molecules. The tetrahedral molecule methane, CH_4, is a good example of a spherical top and we can see, from Fig. 4.11, that its polarizability ellipsoid is spherical. During the vibration known as the symmetric stretch all four C—H distances increase and diminish in phase so that the polarizability ellipsoid contracts and expands but *remains spherical.* For this reason the motion is often referred to as the "breathing frequency"; it is plainly Raman active.

Let us now consider a beam of unpolarized radiation falling on this molecule, and let us designate the direction of this exciting radiation as the z-axis. Since all diameters of a sphere are equal the molecule is equally polarizable in all directions; hence the induced dipole in the molecule will lie along the direction of greatest electric vector in the exciting radiation, i.e. perpendicular to the direction of propagation. Thus the induced dipole will lie in the xy-plane whatever the plane of the incident radiation. This

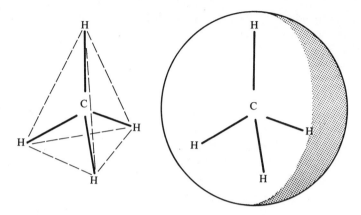

Fig. 4.11: The tetrahedral structure of methane, CH_4, and the spherical polarizability ellipsoid of the molecule.

behaviour is illustrated in Fig. 4.12, where we show an incident beam with its electric vector in (*a*) the vertical (*zy*) plane and (*b*) some other plane making an angle α with the horizontal. In both cases the induced dipole is in the *xy*-plane. A non-polarized incident beam will contain components having all values α.

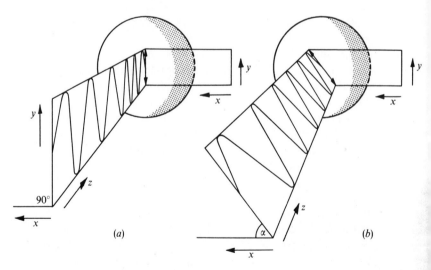

Fig. 4.12: To illustrate the plane polarization of Raman scattering from the symmetric vibration ("breathing vibration") of a spherical top molecule.

To an observer studying the scattered radiation at right angles to the incident beam, i.e. along the x-axis, the oscillating dipole emitting the radiation is confined to the xy-plane—the radiation is plane polarized. When the molecule undergoes the breathing vibration, the polarizability ellipsoid remains spherical and the dipole change remains in the xy-plane. Thus for this vibration the Raman line will be completely plane polarized, and $\rho = 0$, quite irrespective of the nature of the exciting radiation.

We will now consider a less symmetric vibration of this molecule, for example the asymmetric stretching mode (Fig. 4.13(a)) where one C—H bond stretches while the other three contract, and vice versa.

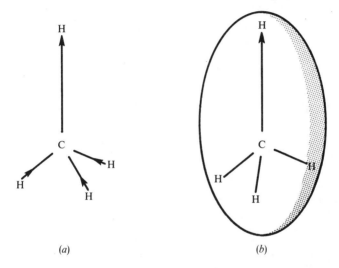

(a)   (b)

Fig. 4.13: An asymmetrical stretching vibration of a spherical top molecule together with the polarizability ellipsoid resulting at the extreme of the motion.

During this vibration the polarizability surface loses its spherical symmetry and becomes ellipsoidal at the extremes of the vibration. One such extreme is shown in Fig. 4.13(b). Now when exciting radiation interacts with the molecule the induced dipole moment will be greatest along the direction of easiest polarizability, i.e. along one of the *minor* axes of the ellipsoid. In a sample of molecules these axes will be oriented in random directions to the incident radiation, so now, because of the lack of spherical symmetry in the

10

vibration, the induced dipole will be randomly oriented and the observed Raman line will be depolarized.

Thus we have immediately a method of assigning some observed Raman lines to their appropriate molecular vibrations—in the case of methane the totally symmetric vibration gives rise to a completely polarized Raman line whereas the non-symmetric vibrations give depolarized lines. The degree of polarization of spectral lines can be readily estimated by noting how the intensity of each line varies when a piece of polaroid or other analyser is put into the scattered radiation first with its polarizing axis parallel to the xy-plane (where z is defined by the direction of the incident beam) and secondly perpendicular to this plane.

4.3. Extension to Other Types of Molecule. Precise calculation, rather than the somewhat pictorial argument used above, shows that Raman scattering may be to some extent polarized when emitted by molecules with less symmetry than the tetrahedral ones. In general it can be stated that a symmetric vibration gives rise to a polarized or partially polarized Raman line while a non-symmetric vibration gives a depolarized line. Theoretically, if the degree of depolarization, ρ, is less than or equal to $\frac{6}{7}$, then the vibration concerned is symmetric and the Raman line is described as "polarized", while if $\rho > \frac{6}{7}$ the line is "depolarized" and the vibration non-symmetric. If we can speak loosely of molecules with increasing symmetry—e.g. linear molecules are less symmetric than symmetric tops which, in turn, are less symmetric than spherical tops—then the higher the molecular symmetry the smaller will be the degree of depolarization of the Raman line for a particular type of vibration.

We can see the usefulness of polarization measurements by considering a simple example. The molecule nitrous oxide has the formula N_2O. Knowing nothing about the structure of this molecule we might turn for help to its infra-red and Raman spectra. The strongest lines in these spectra are collected in Table 4.3 together with their band contours (infra-red) and state of polarization (Raman):

TABLE 4.3: Infra-Red and Raman Spectra of Nitrous Oxide

$\bar{v}$ (cm^{-1})	Infra-Red	Raman
589	Strong; PQR contour	—
1285	Very strong; PR contour	Very strong; polarized
2224	Very strong; PR contour	Strong; depolarized

The data tells us immediately that the molecule has no centre of symmetry (Raman and infra-red lines occur at the same wavenumber) and so the structure is *not* N—O—N. The fact that some infra-red bands have *PR* contour indicates that the molecule is linear, however, so we are led to the conclusion that the structure is N—N—O. Such a molecule should have $3N - 5 = 4$ fundamental modes but two of these (the bending modes) will be degenerate; all three different fundamental frequencies should be both infra-red and Raman active but we note that the perpendicular infra-red band (plainly to be associated with the bending mode) does not appear in the Raman. This accords with expectations—bending modes are often weak and even unobservable in the Raman.

We are left with the assignment of the 1285 and 2224 cm^{-1} bands to the symmetric and asymmetric stretching modes. Both infra-red bands have the same *PR* (parallel) contour, but we note that only the 1285 cm^{-1} is Raman polarized. This, then, we assign to the symmetric mode, leaving the 2224 cm^{-1} band as the asymmetric.

The analysis would not normally rest there. The overtone and combination bands would also be studied to ensure that their activities and contours are in agreement with the molecular model proposed; the fine structure of the infra-red bands also support the structure; and finally isotopic substitution leads to changes in vibrational frequencies in excellent agreement with the model and assignments.

In this rather simple case polarization data was hardly essential to the analysis, but certainly useful. In more complicated molecules it can give very valuable information indeed.

5. Structure Determination from Raman and Infra-Red Spectroscopy

In this section we shall discuss some examples of the combined use of Raman and infra-red spectroscopy to determine the shape of some simple molecules. The discussion must necessarily be limited and the molecules considered (CO_2, N_2O, SO_2; NO_3^-, ClO_3^- and ClF_3) have been chosen to illustrate the principles used; extension to other molecular types should be obvious.

Dealing first with the triatomic AB_2 molecules, the questions to be decided are whether each molecule is linear or not and, if linear, whether it is symmetrical (B—A—B) or asymmetrical (B—B—A). In the case of carbon dioxide and nitrous oxide, both molecules give rise to some infra-red bands with PR contours; they must, therefore, be linear. The mutual exclusion rule (cf. Section 3.2) shows that CO_2 has a centre of symmetry (O—C—O) while N_2O has not (N—N—O), since only the latter has bands common to both its infra-red and Raman spectra. Thus the structures of these molecules are completely determined.

The infra-red and Raman absorptions of SO_2 are collected in Table 4.4.

TABLE 4.4: Infra-Red and Raman Bands of Sulphur Dioxide

Wavenumber	Infra-Red Contours	Raman
519	∥ type band	polarized
1151	∥ type band	polarized
1361	⊥ type band	depolarized

We see immediately that the molecule has no centre of symmetry, since all three fundamentals are both Raman and infra-red active. In the infra-red all three bands show very complicated rotational fine structure, and it is evident that the molecule is non-linear—no band shows the simple PR structure of, say, carbon dioxide. The molecule has, then, the bent $\overset{S}{\diagup \diagdown}_{O \quad O}$ shape.

The AB_3 type molecules require rather more discussion. In general we would expect $3N-6=6$ fundamental vibrations for these 4-atomic molecules. However, if the molecular shape has some symmetry this number will be reduced by degeneracy. In particular, for the symmetric planar and symmetric pyramidal shapes, one stretching mode and one angle deformation mode are each doubly degenerate and so only four different fundamental frequencies should be observed. These are sketched in Table 4.5 where their various activities and band contours or polarizations are also collected. Both molecular shapes are in fact symmetric tops with the main (three-fold) axis passing through atom A perpendicular to the B_3 plane. It is with respect to this axis that the vibrations can be described as ∥ or ⊥. The symmetric modes of vibration are parallel and Raman polarized while the asymmetric are perpendicular and depolarized. All the vibrations of the pyramidal molecule change both the dipole moment and the polarizability, hence all are both infra-red and Raman active. The symmetric stretching mode (v_1) of the planar molecule, however, leaves the dipole moment unchanged (it remains zero throughout) and so is infra-red inactive, while the symmetric bending mode does not change the polarizability (cf. the discussion of the bending mode of CO_2 in Section 3.1) and so v_2 is Raman inactive for planar AB_3.

The overall pattern of the spectra, then, should be as follows:

Planar AB_3: 1 vibration Raman active only (v_1)

1 infra-red active only (v_2)

2 vibrations both Raman and infra-red active (v_3, v_4).

TABLE 4.5: *Activities of Vibrations of Planar and Pyramidal AB_3 Molecules*

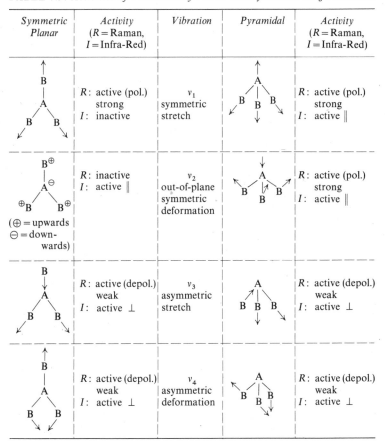

Symmetric Planar	Activity (R = Raman, I = Infra-Red)	Vibration	Pyramidal	Activity (R = Raman, I = Infra-Red)
	R: active (pol.) strong I: inactive	v_1 symmetric stretch		R: active (pol.) strong I: active ∥
(⊕ = upwards ⊖ = downwards)	R: inactive I: active ∥	v_2 out-of-plane symmetric deformation		R: active (pol.) strong I: active ∥
	R: active (depol.) weak I: active ⊥	v_3 asymmetric stretch		R: active (depol.) weak I: active ⊥
	R: active (depol.) weak I: active ⊥	v_4 asymmetric deformation		R: active (depol.) weak I: active ⊥

Pyramidal AB_3: All four vibrations both Raman and infra-red active.

Non-symmetric AB_3: Possibly more than four different fundamental frequencies.

With this pattern in mind we can consider the spectra of $NO_3{}^-$ and $ClO_3{}^-$ ions. The spectroscopic data are summarized in Table 4.6.

TABLE 4.6: Infra-Red and Raman Spectra of NO_3^- and ClO_3^-

Nitrate Ion ($NO_3{}^-$)			Chlorate Ion ($ClO_3{}^-$)		
Raman (cm^{-1})	Infra-Red (cm^{-1})	Assignment	Raman (cm^{-1})	Infra-Red (cm^{-1})	Assignment
690	680 $\perp$	v_4	450 (depol.)	434 $\perp$	v_4
—	830 $\parallel$	v_2	610 (pol.)	624 $\parallel$	v_2
1049	—	v_1	940 (depol.)	950 $\perp$	v_3
1355	1350 $\perp$	v_3	982 (pol.)	994 $\parallel$	v_1

Without considering any assignment of the various absorption bands to particular vibrations, we can see immediately that the nitrate ion fits the expected pattern for a planar system, while the chlorate ion is pyramidal. Detailed assignments follow by comparison with Table 4.5. Thus for the nitrate ion, the band which is Raman active only is obviously v_1 while that which appears only in the infra-red is v_2. If we make the very reasonable assumption that stretching frequencies are larger than bending, then the assignment of v_3 and v_4 is self-evident. This same assumption, coupled with polarization and band contour data, gives the assignment shown in the table for the chlorate ion.

Finally we consider the spectroscopic data for ClF_3. This is found to have no less than *six* strong (and hence fundamental) infra-red absorptions, some of which also occur in the Raman. We know immediately, then, that the molecule is neither symmetric planar nor pyramidal. A complete analysis is not possible from the Raman and infra-red spectra alone, but the use of microwave spectroscopy shows that the molecule is T-shaped with bond angles of nearly 90°.

6. Techniques and Instrumentation

Raman spectroscopy is essentially emission spectroscopy; the scattered radiation from a sample can be examined in the usual way by passage through a dispersing system (e.g. a prism) and subsequent detection and recording of the scattered intensity at different frequencies.   Figure 4.14 shows, very schematically, the essential components of a Raman spectrometer.   We can deal briefly with each component in turn.

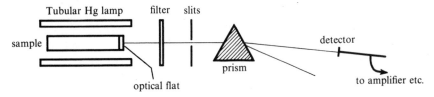

Fig. 4.14: Schematic diagram of a Raman spectrometer.

(i) *The Exciting Source.*   The main requirement of a Raman source is that it be very monochromatic and intense; the most usual source is a mercury lamp which emits suitable radiation at wavelengths of 2536·5 Å and 4358·3 Å.   Of course both these lines, together with other weaker ones, fall on the sample simultaneously and hence each acts as an exciting line during an experiment. However, the lines are sufficiently well separated for the resulting Raman lines not to overlap and no confusion is caused.   The lamp is usually constructed in tube form so that the sample, when placed along the centre of the lamp, is equally and strongly irradiated from all directions.

Recently experiments have been conducted using a LASER source.   These devices are characterized by an extremely sharp and intense emission, ideally suited to Raman work.   A considerable amount of very high resolution and high sensitivity work is expected to result from their use.

(ii) *The Sample Tube and Sample.*   The sample tube is made of glass or quartz and constructed so as to fit snugly into the ring of exciting lamps.   It is usually 20–30 cm long and 1–2 cm in diameter and hence a considerable volume of sample is required.   One end of the tube is blackened or tapered so as to prevent multiple reflections, while the other is made optically flat so that the scattered radiation suffers no distortion on exit.

The sample, which must be most carefully filtered to remove all extraneous scattering centres, is usually a pure liquid since this gives the maximum molecular density and enhances the inherently weak Raman scattering. Gases or dilute solutions can be studied, however, and it is to be expected that their study will increase with the introduction of LASER techniques mentioned above. Solids can also be studied, although only with difficulty.

(*iii*) *The Filter and Modulator.* Since Rayleigh scattering is very much more intense than Raman, it is helpful if the intensity of the Rayleigh line can be lessened before passing on to the detector. In the first place, with a photoelectric recorder, the Rayleigh scattering observed directly would certainly damage the photocell, and secondly the wings of the intense Rayleigh line effectively mask any weak Raman scattering within 100 cm^{-1} or so of the centre of the exciting line. Various filters have been designed to reduce the Rayleigh line as far as possible, but the most efficient is a vessel containing mercury vapour—if the exciting line is the 2537 Å emission of a mercury lamp. The vapour (being in the ground state) will only absorb radiation of its own ground to first excited state frequency and thus will diminish the Rayleigh line without affecting the Raman spectrum.

When a photoelectric detector is used a modulator is often put into the Raman beam at the entrance to the dispersing prism. This, which may be simply a rotating sector, modulates the spectrum with a certain frequency to which the amplifying equipment may be tuned. As discussed earlier, this can produce a considerable improvement in the signal-to-noise ratio.

(*iv*) *The Detector.* For reasonably intense Raman lines (e.g. from a pure liquid) this is most conveniently a photoelectric recording device such as a photomultiplier tube. Using this a complete spectrum can be obtained and recorded directly on to chart paper. The spectrum of chloroform shown in Fig. 4.9(*a*) was obtained in this way.

If the Raman lines are weak, however, either inherently or because the sample is dilute, it is often preferable to record the spectrum on a photographic plate, since the plate has the advantage that the weakest of lines can be detected by a sufficiently long exposure. Exposure times of several days are not uncommon.

7. Summary

Raman spectroscopy, we have seen, gives essentially the same sort of information about molecular structure and dimensions as do infra-red and microwave spectroscopy. It suffers under the major disadvantage that the weakness of its spectra leads to resolution distinctly inferior to both these techniques and so the finer details of rotational fine structure, for example, are only observed in favourable cases.   However, it possesses some advantages.

From the point of view of technique, it has essentially transferred microwave and infra-red studies into the visible region of the spectrum where photographic methods have been particularly well developed for the study of very weak effects.   Furthermore, in vibrational studies, the usual lower limit for infra-red methods is some 400 cm^{-1}—although special instruments do go lower than this, they are relatively expensive.   Raman techniques using a mercury lamp source can observe vibrational transitions as low as 100 cm^{-1} and the use of LASERS, with their sharper exciting lines, will probably reduce this limit considerably.

Techniques apart, however, Raman spectroscopy must not be considered as a poor relation to infra-red and microwave.   By its means, we can, for instance, study the rotations and vibrations of molecules such as O_2 or H_2 which are quite inaccessible to infra-red or microwave techniques; or we can observe the symmetric vibrations of molecules, such as CO_2, which produce no dipole change and therefore no infra-red spectrum.   Indeed, we have seen that for centro-symmetric molecules Raman and infra-red studies are exactly complementary.   For other molecules too, Raman techniques yield data which cannot be obtained otherwise, or which help in the interpretation of other data (e.g. the degree of depolarization of a Raman line assists the assignment of a corresponding infra-red vibration).

The comparative brevity of this chapter must not be taken as a measure of the importance of Raman spectroscopy.   The brevity has come about because much of the detailed algebra of infra-red and microwave spectra could be taken over directly into the Raman. No spectroscopist should be content with an analysis based on the examination of an infra-red spectrum alone at least for symmetrical molecules; it should always be studied, wherever possible, in conjunction with the corresponding Raman spectrum.

Bibliography

BERL, W. G. (Ed.): *Physical Methods in Chemical Analysis*, Vol. I, 2nd edn., Academic Press, 1960

HERZBERG, G.: *Molecular Spectra and Molecular Structure*, Vol. 1, *Spectra of Diatomic Molecules*, 2nd edn., Van Nostrand, 1950

HERZBERG, G.: *Molecular Spectra and Molecular Structure*, Vol. 2, *Infra-red and Raman Spectra of Polyatomic Molecules*, Van Nostrand, 1945

KING, G. W.: *Spectroscopy and Molecular Structure*, Holt, Rinehart and Winston Inc., 1964

NACHOD, F. C. and W. D. PHILLIPS: *Determination of Organic Structures by Physical Methods*, Vol. 2, Academic Press, 1962

WALKER, S. and H. STRAW: *Spectroscopy*, Vol. II, Chapman & Hall, 1962

WIBERLEY, S. E., N. B. COLTHUP and L. H. DALY: *Introduction to Infra-red and Raman Spectroscopy*, Academic Press, 1964

CHAPTER 5

ELECTRONIC SPECTROSCOPY
OF ATOMS

1. The Structure of Atoms

1.1. Electronic Wave Functions. It is well-known that an atom consists of a central, positively charged nucleus, which contributes nearly all the mass to the system, surrounded by negatively charged electrons in sufficient number to balance the nuclear charge. Hydrogen, the smallest and simplest atom, has a nuclear charge of $+1$ units (where the unit is the electronic charge, $4{\cdot}80 \times 10^{-10}$ e.s.u.) and one electron; each succeeding atom increases the nuclear charge and electron total by unity, up to atoms with 100 or more electrons.

Modern theories have long ceased to regard the electron as a particle which obeys the laws of classical mechanics applicable to massive, everyday objects; instead, in common with all entities of sub-atomic size, we consider that it obeys the laws of quantum mechanics (or wave mechanics) as embodied in the Schrödinger wave equation. In principle, this equation may be used to determine many things: e.g. the way in which electrons group themselves about a nucleus when forming an atom; the energy which each electron may have; the way in which it can undergo transitions between energy states, etc. In practice the application of the Schrödinger equation to these problems presents difficulties which can only be overcome in the case of the simplest atoms or by the use of gross approximations. Here, however, we shall be concerned only with the results obtained—and then only in qualitative terms —rather than the mathematical theory of the process.

The Schrödinger theory can be used to predict the *probability* of an electron with a particular energy being at a particular point in space, and it expresses this probability in terms of a very important algebraical expression called the *wave-function* of the electron. The wave function is given the Greek symbol ψ. Quite simply, the probability of finding an electron, whose wave-function is ψ, within

145

unit volume at a given point in space is proportional to the value of ψ^2 at that point:

$$\text{relative probability density} = \psi^2. \qquad (5.1)$$

Let us see what this means. Electronic wave functions consist of three elements: (i) some fundamental *physical constants* (π, h, c, m, e, etc.—where m and e are the mass and charge, respectively, of the electron); (ii) *parameters* peculiar to the system under discussion— e.g. for atoms, distance from the nucleus, either radially (r) or along some co-ordinate axes (x, y, z); and (iii) one or more *quantum numbers*. These latter are by no means arbitrarily introduced into the problem in order to make the predictions match experiment; they *belong* to the solution of the Schrödinger equation in the sense that ψ represents a sensible physical situation *only if* the quantum numbers have certain values.

As an example we may quote here the expression for a set of wave functions, ψ_n, which are solutions to the Schrödinger equation for the hydrogen atom:

$$\psi_n = f\left(\frac{r}{a_0}\right) \exp\left(-\frac{r}{na_0}\right) \qquad (5.2)$$

where $a_0 = h^2/4\pi^2 me^2$, r is radial distance from the nucleus, $f(r/a_0)$ is a power series of degree $(n-1)$ in r/a_0, and n is the *principal quantum number*, which can have only integral values, 1, 2, 3, ..., ∞. The constant a_0 has dimensions of length (and is, in fact, about 0.5×10^{-8} cm), and so the quantity r/a_0 is a pure number. Thus for particular values of r and n, ψ_n and ψ_n^2 are also simply numbers, and ψ_n^2 represents the probability of finding the electron at our chosen distance r from the nucleus when it is in the state represented by the given n-value.

It is found that the electronic wave functions of all atoms require the introduction of only four quantum numbers. We shall describe these briefly here, leaving a more thorough discussion to later sections.

1.2. The Shape of Atomic Orbitals; Atomic Quantum Numbers. Table 5.1 lists the four quantum numbers, gives the allowed values of each and states what is the function of each. The *principal quantum number*, as stated earlier, can take integral values from 1 to infinity. It governs the energy of the electron mainly (although we shall see later the other quantum numbers also affect this energy to some extent).

TABLE 5.1: The Atomic Quantum Numbers

Quantum No.	Allowed Values	Function
Principal, n	$1, 2, 3, \ldots$	Governs the energy and size of the orbital
Azimuthal, l	$(n-1), (n-2), \ldots, 0$	Governs the shape of the orbital and the electronic angular momentum
Magnetic, m	$\pm l, \pm(l-1), \ldots, 0$	Governs the direction of an orbital and the electrons' behaviour in a magnetic field
Spin, s	$+\frac{1}{2}$	Governs the axial angular momentum of the electron

The table states that n also governs the size of the electronic *orbital*; this latter is a term used to represent the space within which an electron can move according to the Schrödinger theory—it corresponds approximately to the earlier idea of Bohr that electrons move in circular or elliptical *orbits* like planets round a sun. Energy and size of the orbital are connected in that the smaller the orbital the closer to the nucleus the electron will be and hence the more firmly bound.

The *azimuthal* (or *orbital*) *quantum number l*, also has integral values only, but these must be less than n. Thus for $n=3$, l can be 2, 1 or 0. It governs the shape of the orbital (cf. Fig. 5.1) and the angular momentum of the electron as it circulates about the nucleus in its orbital.

The *magnetic quantum number m*, takes integral values which depend on l. Thus for $l=2$, m can be $+2$, $+1$, 0, -1 or -2; in general there are $2l+1$ values of m. Besides denoting the behaviour of electrons in orbitals when the atom is placed in a magnetic field, the m quantum number can also be used to specify the direction of a particular orbital.

The *spin quantum number s*, is of magnitude $+\frac{1}{2}$ only (but cf. Section 2.2). It measures the spin angular momentum which the electron possesses whether it is present in an atom or in free space.

Since wave functions represent only a probability distribution of an electron it is difficult to define precisely the shape and size of an orbital. From equation (5.2) we see that even at very large values of r, ψ_n and hence ψ_n^2, still has a value, even though small. Thus an orbital tails off to infinity (although, because of the smallness of a_0, "infinity" on the atomic scale might be taken as 10^{-4} or 10^{-5} cm) in all directions. However, the difficulty can be overcome if we agree to draw a three-dimensional shape *within* which the electron spends, say, 95% or some other fraction, of its time. This can be taken as the effective boundary of the electron's domain and it can be called *the orbital*.

Considering still the wave-function of equation (5.2) we see that the corresponding orbital must be spherical; for any given distance r from the nucleus, ψ_n has the same value irrespective of direction. Thus the 95% boundary will be spherical. For larger n the function tails off less rapidly with distance and so the electron can spend

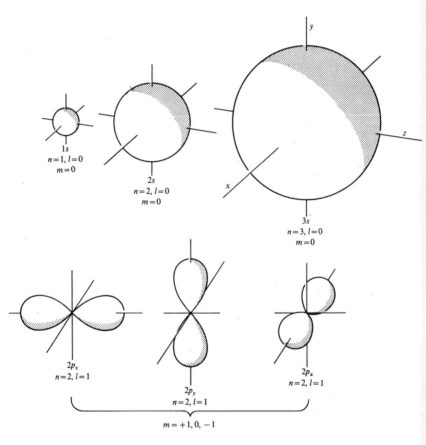

Fig. 5.1: *Some of the electronic orbitals which may be occupied by the electron in a hydrogen atom.*

proportionately more of its time further from the nucleus; thus the 95% sphere will increase in size with n. We have drawn the cases $n = 1$, 2 and 3 at the top of Fig. 5.1. These spherical orbitals, it so happens, are associated with an l-value of zero (and hence $m = 0$) and they are referred to as s *orbitals*. (Although it is perhaps helpful to connect s with "spherical"; in fact the label arose historically because of the alleged particular *sharpness* of spectral lines arising from transition of electrons occupying s orbitals; in fact the connection which should be remembered is between s orbitals and $l = 0$.) The s orbitals are labelled according to their n quantum numbers: $1s, 2s, \ldots, ns$.

Orbitals with $l = 1$ (and hence $n \geqslant 2$) also arise as solutions to the Schrödinger equation for the hydrogen atom. These are twin-lobed and have the approximate shape shown for $n = 2$ in the lower half of Fig. 5.1. Orbitals with $n = 3$, $l = 1$ are larger but have the same shape. Such orbitals are labelled p (historically their transitions were thought to be "principal") and we see that, for a given n, there are *three* of them, one along each co-ordinate axis. They can be distinguished as np_x, np_y and np_z, if necessary, or they can be specified by attaching to each, one of the three allowed m-values for $l = 1$ (i.e. $m = +1$, 0 and -1).

We can go further: for $l = 2$ (hence $n \geqslant 3$) we have a set of d *orbitals* (historically "diffuse"), and $l = 3$ $(n \geqslant 4)$ f *orbitals* (historically "fundamental"): there are five of the former $(m = \pm 2, \pm 1$ or 0) and seven of the latter $(m = \pm 3, \pm 2, \pm 1$ or 0). Sketches of d orbitals show that they have four lobes, while the f have six, but we shall not attempt to reproduce these here. Orbitals with higher l values, $l = 4, 5, 6, \ldots$, are of less importance and we shall not consider them further; if necessary they are labelled alphabetically after f, i.e. $l = 4$, g; $l = 5$, h, etc.

1.3. The Energies of Atomic Orbitals; Hydrogen Atom Spectrum. However large an atom its electrons take up orbitals of the s, p, d, ... type (according to very specific laws which we shall discuss later) and so the overall shape of each electron's domain is unaltered. The *energy* of each orbital, on the other hand, varies considerably from atom to atom. There are two main contributions to this energy: (*i*) attraction between electrons and nucleus, (*ii*) repulsion between electrons in the same atom. Factor (*ii*) plainly depends on the number of electrons in the atom and hence the orbital energies are quite variable.

We consider first the case of hydrogen in some detail: this is the simplest because, having only one electron, factor (*ii*) is completely absent.   We shall later see how the picture should be modified for larger atoms.

Because of the absence of interelectronic effects all orbitals with the same *n*-value have the same energy in hydrogen.   Thus the 2*s* and 2*p* orbitals, for instance, are degenerate, as are the 3*s*, 3*p* and 3*d*. However the energies of the 2*s*, 3*s*, 4*s*,... orbitals differ considerably.   For the *s* orbitals given by equation (5.2):

$$\psi_{ns} = f\left(\frac{r}{a_0}\right) \exp\left(-\frac{r}{na_0}\right)$$

the Schrödinger equation shows that the energy is:

$$E_n = -\frac{2\pi^2 m e^4}{h^2 n^2} \text{ ergs,}$$

or

$$\varepsilon_n = -\frac{2\pi^2 m e^4}{h^3 c n^2} = -\frac{R}{n^2} \text{ cm}^{-1} \quad (n = 1, 2, 3, \ldots) \tag{5.3}$$

where the fundamental constants have been collected together and given the symbol *R*, called the Rydberg constant.   Since *p*, *d*,... orbitals have the same energies as the corresponding *s* (for hydrogen *only*), equation (5.3) represents *all* the electronic energy levels of this atom.

The lowest value of ε_n is plainly $\varepsilon_n = -R$ cm^{-1} (when $n=1$), and so this represents the most stable (or ground) state. ε_n increases with increasing *n*, reaching a limit, $\varepsilon_n = 0$ for $n = \infty$.   This represents complete removal of the electron from the nucleus, i.e. the state of ionization.   We sketch these energy levels for $n = 1$ to 5 and $l = 0, 1$ and 2 only in Fig. 5.2.   (Some possible transitions, also shown, will be discussed shortly.)   The three *p* states and five *d* states for each *n* are degenerate and not shown separately.

Equation (5.3) and Fig. 5.2, then, represent the energy levels of the atom; in order to discuss the spectra which may arise we need the selection rules governing transitions.   The Schrödinger equation shows these to be:

$$\Delta n = \pm 1, \pm 2, \pm 3, \ldots \quad \text{and} \quad \Delta l = \pm 1 \text{ only.} \tag{5.4}$$

From these selection rules we see immediately that an electron in the ground state (the $1s$) can undergo a transition into any p state:

$$1s \rightarrow np \quad (n \geqslant 2)$$

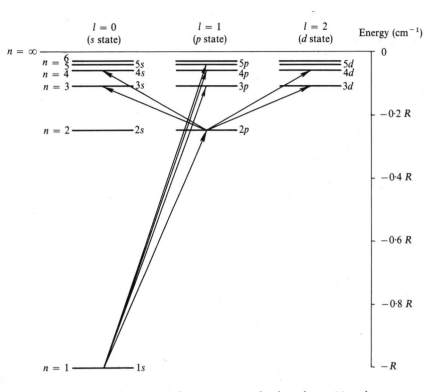

Fig. 5.2: *Some of the lower electronic energy levels and transitions between them for the single electron of the hydrogen atom.*

while a $2p$ electron can have transitions either into an s state or a d state:

$$2p \rightarrow ns \text{ or } nd \quad (n \neq 2).$$

Since s and d orbitals are here degenerate the energy of both these transitions will be identical. These transitions are sketched in Fig. 5.2.

In general an electron in a lower state n'', can undergo a transition into a higher state n', with absorption of energy:

$$\Delta\varepsilon = \varepsilon_{n'} - \varepsilon_{n''} \text{ cm}^{-1}$$

$$\therefore \bar{\nu}_{\text{spect.}} = -\frac{R}{n'^2} - \left(-\frac{R}{n''^2}\right) = R\left\{\frac{1}{n''^2} - \frac{1}{n'^2}\right\} \text{ cm}^{-1}. \quad (5.5)$$

An identical spectral line will be produced in emission if the electron falls from state n' to state n''. In both cases l must change by unity. Let us consider a few of these transitions, restricting ourselves to absorption for simplicity.

Transitions $1s \to n'p$, $n' = 2, 3, 4, \ldots$ For these

$$\bar{\nu}_{\text{Lyman}} = R\left\{\frac{1}{1} - \frac{1}{n'^2}\right\} = R - \frac{R}{n'^2} \text{ cm}^{-1}$$

$$= \frac{3R}{4}, \frac{8R}{9}, \frac{15R}{16}, \frac{24R}{25}, \ldots \text{ cm}^{-1} \quad (\text{for } n' = 2, 3, 4, 5, \ldots).$$

Hence we expect a series of lines at the wavenumbers given above. Just such a series is indeed observed in the atomic hydrogen spectrum, and it is called the Lyman series after its discoverer. The appearance of this spectrum is sketched in Fig. 5.3 together with a scale in units of R and in wavenumbers. We can see that the spectrum converges to the point R cm^{-1}, and from the observed spectrum the very precise value $R = 109{,}677 \cdot 581$ cm^{-1} is obtained. This convergence limit, which arises when $n' = \infty$, is shown dotted on the figure. It plainly represents complete removal of the

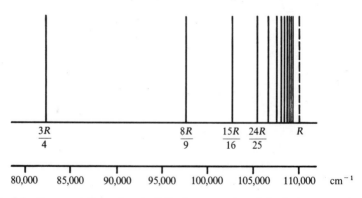

Fig. 5.3: *Representation of part of the Lyman series of the hydrogen atom showing the convergence (ionization) point.*

electron—i.e. ionization—and the energy required to ionize the atom is given, in cm^{-1}, by the value of R. Using the conversion factor $1\ cm^{-1} = 1\cdot24 \times 10^{-4}$ eV, we have a very precise measure of the *ionization potential* from the ground $(1s)$ state: 13·595 eV.

Another set of transitions arises from an electron initially in the $2s$ or $2p$ states: $2s \rightarrow n'p$ or $2p \rightarrow n's, n'd$. For these we write:

$$\bar{\nu}_{Balmer} = R\left\{\frac{1}{4} - \frac{1}{n'^2}\right\} cm^{-1}$$

$$= \frac{5R}{36}, \frac{3R}{16}, \frac{21R}{100}, \ldots\ cm^{-1} \quad (\text{for } n' = 3, 4, 5, \ldots).$$

Thus we expect another series of lines converging to $\frac{1}{4}R\ cm^{-1}$ $(n' = \infty)$; this series, called the Balmer series after its discoverer, is observed and the value of $\frac{1}{4}R$ obtained from its convergence limit— which represents the ionization potential from the first excited state—is in excellent agreement with the value of R from the Lyman series.

Other similar line series (called the Paschen, Brackett, Pfund, etc., series) are observed for $n'' = 3, 4, 5, \ldots$; indeed these spectra were observed long before the modern theory of atomic structure had been developed. The spectral lines were correlated empirically by Rydberg, and he showed that an equation of the form given in equation (5.5) described the wavenumbers of each. It is after him that the Rydberg constant is named.

It should be mentioned that each line series discussed above shows a *continuous absorption* or *emission* to high wavenumbers of the convergence limits. The convergence limit represents the situation where the atomic electron has absorbed just sufficient energy from radiation to escape from the nucleus with zero velocity. It can, however, absorb more energy than this and hence escape with higher velocities and since the kinetic energy of an electron moving in free space is *not* quantized, *any* energy above the ionization energy can be absorbed. Hence the spectrum in this region is continuous.

This completes our discussion of what might be termed the coarse structure of the hydrogen atom spectrum. In order to consider the fine structure we need to know how the other quantum numbers, besides n, affect the electronic energy levels.

2. Electronic Angular Momentum

2.1. Orbital Angular Momentum. An electron moving in its orbital about a nucleus possesses *orbital* angular momentum, a measure of which is given by the l value corresponding to the orbital. This momentum is, of course, quantized, and it is usually expressed in terms of the unit $h/2\pi$, where h is Planck's constant. We may write:

$$\text{orbital angular momentum} = \sqrt{l(l+1)}\cdot\frac{h}{2\pi} = \sqrt{l(l+1)} \text{ a.m. units.}$$

(5.6)

Now angular momentum is a *vector* quantity, by which we mean that its *direction* is important as well as its magnitude—the axis of a spinning top, for instance, points in a particular direction. Conventionally, vectors may be represented by arrows, and the angular momentum vector is represented by an arrow based at the centre of the top, along the top axis, and of length proportional to the magnitude of the angular momentum. Such an arrow can lie in two different directions, at 180° to each other; these directions are associated, depending on the sign convention used, with clockwise and anticlockwise rotations of the top. Mathematically we can ignore the spinning body and deal merely with the properties of the arrow.

It is usual to distinguish vector quantities by the use of *bold face* type and we shall accordingly represent orbital angular momentum by the symbol **l** where:

$$\mathbf{l} = \sqrt{l(l+1)} \text{ a.m. units.}$$

(5.7)

In this equation l is always zero or positive and hence so is **l**. Since **l** and l are so closely connected they are often loosely used interchangeably: thus we speak of an electron having "an angular momentum of 2" when we strictly mean that $l=2$ and $\mathbf{l}=\sqrt{2\times 3}=$ 2·44 units.

We might at first think that the angular momentum vector of an electron could point in an infinite number of different directions. This, however, would be to reckon without the quantum theory. In fact, once a reference direction has been specified (and this may be done in many ways, either externally, such as by applying an electric or magnetic field, or internally, perhaps in terms of the

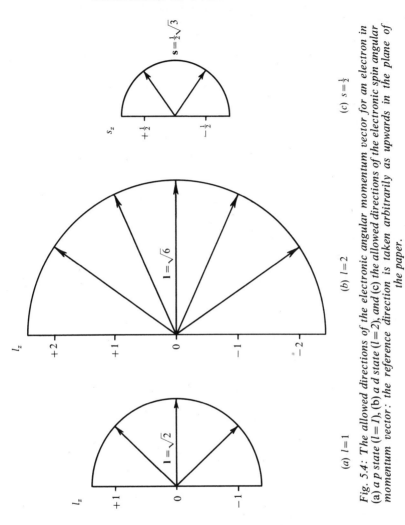

Fig. 5.4: The allowed directions of the electronic angular momentum vector for an electron in (a) a p state (l=1), (b) a d state (l=2), and (c) the allowed directions of the electronic spin angular momentum vector: the reference direction is taken arbitrarily as upwards in the plane of the paper.

angular momentum vector of one particular electron), the angular momentum vector can point *only so as to have integral components along the reference direction.* Figure 5.4(a) and (b) show the situation for an electron with $l=1$ and $l=2$ respectively (i.e. a p and a d electron). The reference direction, here from top to bottom of the figure, is usually taken as the z-direction, and we can hence write the *components of* l in this direction as l_z. In general we see that l_z

has values:

$$l_z = l, l-1, \ldots, 0, \ldots, -(l-1), -l \qquad (5.8)$$

and that there are $2l+1$ values of l_z for a given l. Plainly l_z is to be identified with the magnetic quantum number m introduced in Section 1.2:

$$l_z \equiv m$$

and this justifies our previous assertion that m governs essentially the *direction* of an orbital.

Before proceeding further let us reiterate the distinction between l, **l** and l_z. The quantum number l is an integer, positive or zero, representing the state of an electron in an atom and determining its orbital angular momentum. The vector **l** designates the magnitude and direction of this momentum as shown by the vector arrows of Fig. 5.4. When expressed in units of $h/2\pi$, **l** is numerically equal to $\sqrt{l(l+1)}$. Once a reference direction is specified (and this is often quite arbitrary) **l** can point only so as to have integral components l_z along that direction.

Usually the orbital energy of the electron depends only on the magnitude and not the direction of its angular momentum; thus the $2l+1$ values of l_z are all *degenerate*. But we should note that it is possible to *lift the degeneracy* (cf. Section 5) so that levels with different l_z have different energy.

2.2. Electron Spin Angular Momentum. Every electron in an atom can be considered to be spinning about an axis as well as orbiting about the nucleus. Its spin motion is designated by the *spin quantum number s*, which can be shown to have a value of $\frac{1}{2}$ only. Thus the spin angular momentum **s** is given by:

$$\mathbf{s} = \sqrt{s(s+1)}\,\frac{h}{2\pi} = \sqrt{\frac{1}{2} \times \frac{3}{2}} \text{ a.m. units,}$$

$$= \tfrac{1}{2}\sqrt{3} \text{ a.m. units.} \qquad (5.9)$$

The quantization law for spin momentum is that the vector can point so as to have *half-integral* components in the reference direction, i.e. $s_z = +\frac{1}{2}$ or $-\frac{1}{2}$. The two (i.e. $2s+1$) allowed directions are shown in Fig. 5.4(c); they are normally degenerate.

2.3. Total Electronic Angular Momentum. We now need to discover some means whereby the orbital and spin contributions to the electronic angular momentum may be combined. Formally we can write:

$$\mathbf{j} = \mathbf{l} + \mathbf{s}, \tag{5.10}$$

where $\mathbf{j}$ is the *total angular momentum*. Since $\mathbf{l}$ and $\mathbf{s}$ are vectors, equation (5.10) must be taken to imply *vector addition*. Also formally, we can express $\mathbf{j}$ in terms of a total angular momentum quantum number j:

$$\mathbf{j} = \sqrt{j(j+1)}\,\frac{h}{2\pi} = \sqrt{j(j+1)} \text{ a.m. units}, \tag{5.11}$$

where j is *half-integral* (since s is half integral for a one-electron atom), and a quantal law applies equally to $\mathbf{j}$ as to $\mathbf{l}$ and $\mathbf{s}$: $\mathbf{j}$ *can have z-components which are half-integral only*, i.e.

$$j_z = \pm j, \pm(j-1), \pm(j-2), \ldots, \tfrac{1}{2}. \tag{5.12}$$

There are two methods by which we can deduce the various allowed values of $\mathbf{j}$ for particular $\mathbf{l}$ and $\mathbf{s}$ values. We shall consider them both briefly.

(*i*) Vector Summation. In ordinary mechanics two forces in different directions may be added by a graphical method in which vector arrows are drawn to represent the magnitude and direction of the forces, the "parallelogram is completed" and the magnitude and direction of the resultant given by the diagonal of the parallelogram. Exactly the same method can be used to find the resultant ($\mathbf{j}$) of the vectors $\mathbf{l}$ and $\mathbf{s}$. The important difference is that quantum mechanical laws restrict the angle between $\mathbf{l}$ and $\mathbf{s}$ to values such that $\mathbf{j}$ is given by equation (5.11) with *half-integral j*. Thus $\mathbf{j}$ can take values

$$\tfrac{1}{2}\sqrt{3}, \tfrac{1}{2}\sqrt{15}, \tfrac{1}{2}\sqrt{35}, \ldots \text{ corresponding to } j = \tfrac{1}{2}, \tfrac{3}{2}, \tfrac{5}{2}, \ldots$$

The method is illustrated in Fig. 5.5(*a*) and (*b*) for the case $l=1$ (i.e. $\mathbf{l}=\sqrt{2}$) and $s=\tfrac{1}{2}$ ($\mathbf{s}=\tfrac{1}{2}\sqrt{3}$). In (*a*) the summation yields $\mathbf{j}=\tfrac{1}{2}\sqrt{15}$, which corresponds to a j-value of $\tfrac{3}{2}$, while in (*b*) $\mathbf{j}=\tfrac{1}{2}\sqrt{3}$ or $j=\tfrac{1}{2}$. Construction or calculation shows that $\mathbf{l}$ and $\mathbf{s}$ may not be combined in any other way to give an allowed value of $\mathbf{j}$.

Note that we can get exactly the same answer by summing the *quantum numbers l* and s to get the *quantum number j*. In this example $l=1$, $s=\tfrac{1}{2}$, and hence:

$$j = l+s = \tfrac{3}{2}, \quad \text{or} \quad j = l-s = \tfrac{1}{2}.$$

This simple approach, although adequate for systems with one electron only, is not readily extended to multi-electron systems. For these we must use the rather more fundamental method outlined below.

(*ii*) Summation of *z*-Components. If the components along a common direction of two vectors are added, the summation yields the component in that direction of their resultant. We have seen

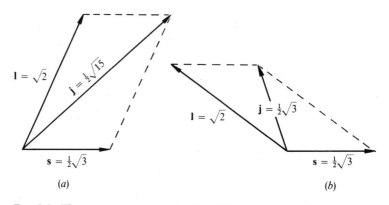

Fig. 5.5: *The two energy states having different total angular momentum which can arise as a result of the vector addition of* $\mathbf{l} = \sqrt{2}$ *and* $\mathbf{s} = \frac{1}{2}\sqrt{3}$.

(cf. equation (5.8)) that the *z*-components of $l=1$ are ± 1 and 0, while those of $s=\frac{1}{2}$ are $\pm\frac{1}{2}$ only. Taking all possible sums of these quantities we have:

$$j_z = l_z + s_z \qquad (5.13)$$

$$\therefore \; j_z = 1+\tfrac{1}{2}, \, 1-\tfrac{1}{2}, \, 0+\tfrac{1}{2}, \, 0-\tfrac{1}{2}, \, -1+\tfrac{1}{2}, \, -1-\tfrac{1}{2};$$

$$= \tfrac{3}{2}, \, \tfrac{1}{2}, \, \tfrac{1}{2}, \, -\tfrac{1}{2}, \, -\tfrac{1}{2}, \, -\tfrac{3}{2}.$$

In this list of six j_z components, the maximum value is $\frac{3}{2}$, which we know (cf. equation (5.12)) must belong to $j=\frac{3}{2}$. Other components of $j=\frac{3}{2}$ are $\frac{1}{2}$, $-\frac{1}{2}$ and $-\frac{3}{2}$ and, striking these from the above six, we are left with $j_z = +\frac{1}{2}$ and $-\frac{1}{2}$. These values are plainly consistent with $j=\frac{1}{2}$.

Thus all the six components are accounted for if we say that the states $j=\frac{3}{2}$ and $j=\frac{1}{2}$ may be formed from $l=1$ and $s=\frac{1}{2}$. This is, of course, in agreement with the vector summation method.

Both these methods show that for a p electron (i.e. $l=1$), the orbital and spin momenta may be combined to produce a total momentum of $\mathbf{j}=\frac{1}{2}\sqrt{15}$ when $\mathbf{l}$ and $\mathbf{s}$ reinforce (physically we would say that the angular momenta have the same *direction*) or to give $\mathbf{j}=\frac{1}{2}\sqrt{3}$ when $\mathbf{l}$ and $\mathbf{s}$ oppose each other. Thus the total momentum is different in *magnitude* in the two cases and hence we have arrived at two *different energy states* depending on whether $\mathbf{l}$ and $\mathbf{s}$ reinforce or oppose. Both energy states are p states, however (since l is 1 for both), and they may be distinguished by writing the j quantum number value as a subscript to the *state symbol P*, thus $P_{3/2}$ or $P_{1/2}$. (We here use a capital letter for the state of a whole atom and a small letter for the state of an individual electron; in the hydrogen atom, which contains only one electron, the distinction is trivial.) States such as these, split into two energies, are termed *doublet states*; their doublet nature is usually indicated by writing a superscript 2 to the state symbol, thus: $^2P_{3/2}$, $^2P_{1/2}$. The state (or term) symbols produced are to be read "doublet p three halves" or "doublet p one half" respectively.

All other higher l values for the electron will obviously produce doublet states when combined with $s=\frac{1}{2}$; for instance, $l=2, 3, 4, \ldots$ will yield $^2D_{5/2, 3/2}$, $^2F_{7/2, 5/2}$, $^2G_{9/2, 7/2}$, etc. The student should satisfy himself of this, preferably by using the z-component summation method outlined above. There is, however, a slight difficulty with s states ($l=0$). Here, since $l=0$, it can make no contribution to the vector sum, and the only possible resultant is $\mathbf{s}=\frac{1}{2}\sqrt{3}$ or $s=\frac{1}{2}$. Remember $s=-\frac{1}{2}$ is not allowed, since the *quantum number* cannot be negative; it is only the z-component of the vector which can have negative values. Thus for an s electron we would have the state symbol $S_{1/2}$ only. This is nonetheless formally written as a doublet state ($^2S_{1/2}$) where the two components are assumed to be degenerate.

We can now consider the relevance of this discussion to atomic spectroscopy.

2.4. The Fine Structure of the Hydrogen Atom Spectrum. The hydrogen atom contains but one electron and so the coupling of orbital and spin momenta and consequent splitting of energy levels will be exactly as described above. We summarize the essential details of the energy levels in Fig. 5.6. Each level is labelled with its n quantum number on the extreme left, and its j-value on the right;

the l-value is indicated by the term symbols $S, P, D, \ldots$ at the top of each column. The S levels are shown degenerate, as described above, while splitting is indicated for the P and D levels; similar splitting would occur for $F, G, \ldots$ states but these are not shown.

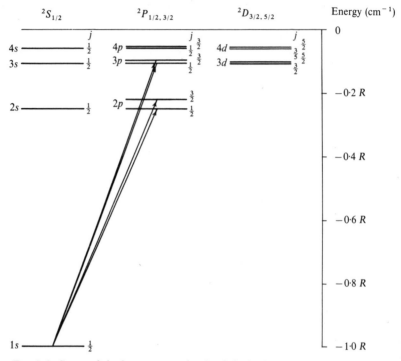

Fig. 5.6: Some of the lower energy levels of the hydrogen atom showing the inclusion of j-splitting; the splitting is greatly exaggerated for clarity.

There is no attempt to show the energy level splitting to scale in this diagram—the separation between levels differing only in j is many thousands of times smaller than the separation between levels of different n. However, we do indicate that the j-splitting *decreases* with increasing n and with increasing l.

The selection rules for n and l are the same as before:

$$\Delta n = \pm 1, \pm 2, \ldots, \qquad \Delta l = \pm 1 \text{ only} \qquad (5.14)$$

but now there is a selection rule for j:

$$\Delta j = 0, \pm 1. \qquad (5.15)$$

These selection rules indicate that transitions are allowed between any S level and any P level:

$$^2S_{1/2} \rightarrow {}^2P_{1/2}: \quad \Delta j = 0$$

$$^2S_{1/2} \rightarrow {}^2P_{3/2}: \quad \Delta j = +1.$$

Thus the spectrum to be expected from the ground ($1s$) state will be identical with the Lyman series (cf. Section 1.3) except that *every line will be a doublet.*   In fact the separation between the lines is too

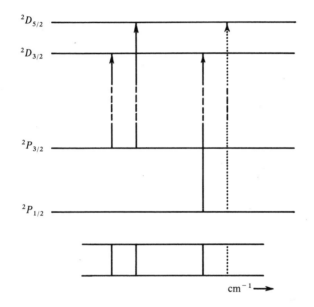

Fig. 5.7: The "compound doublet" spectrum arising as the result of transitions between ^2P and ^2D levels in the hydrogen atom.

small to be readily resolved but we shall shortly consider the spectrum of sodium in which this splitting is easily observed.

Transitions between the ^2P and ^2D states are rather more complex; Fig. 5.7 shows four of the energy levels involved.   Plainly the transition at lowest frequency will be that between the closest pair of levels, the $^2P_{3/2}$ and $^2D_{3/2}$.   This, corresponding to $\Delta j = 0$ is allowed.   The next transition, $^2P_{3/2} \rightarrow {}^2D_{5/2}$ ($\Delta j = +1$) is also allowed and will occur close to the first because the separation between the doublet D states is very small.   Thirdly, and more widely spaced, will be $^2P_{1/2} \rightarrow {}^2D_{3/2}$ ($\Delta j = +1$), but the fourth transition (shown dotted) $^2P_{1/2} \rightarrow {}^2D_{5/2}$, is not allowed since for this $\Delta j = +2$.

Thus the spectrum will consist of the three lines shown at the foot of the figure. This, arising from transitions between doublet levels, is usually referred to as a "compound doublet" spectrum.

We see, then, that the inclusion of coupling between orbital and spin momenta has led to a slight increase in the complexity of the hydrogen spectrum. In practice, the complexity will be observed only in the spectra of heavier atoms, since for them the j-splitting is larger than for hydrogen. In principle, however, all the lines in the hydrogen spectrum should be close doublets, if the transitions involve s levels, or "compound doublets" if s electrons are not involved.

3. Many-Electron Atoms

3.1. The Aufbau Principle. The Schrödinger equation shows that electrons in atoms occupy orbitals of the same type and shape as the $s, p, d, \ldots$ orbitals discussed for the hydrogen atom, but that the energies of these electrons differ markedly from atom to atom. There is no general expression for the energy levels of a many-electron atom comparable to equation (5.3) for hydrogen; each atom must be treated as a special case and its energy levels either tabulated or shown on a diagram similar to Fig. 5.2 or Fig. 5.6.

There are three basic rules, known as the aufbau (or building-up) rules which determine how electrons in large atoms occupy orbitals. These may be summarized as:

1. Pauli's principle: no two electrons in an atom may have the same set of values for n, l, l_z ($\equiv m$) and s_z.

2. Electrons tend to occupy the orbital with lowest energy available.

3. Hund's principle: electrons tend to occupy degenerate orbitals singly with their spins parallel.

Rule 1 effectively limits the number of electrons in each orbital to two. An example may make this clear: we may characterize both an orbital and an electron occupying it by specifying the n, l and m quantum numbers. Thus a $1s$ orbital or $1s$ electron has $n=1$, $l=0$ and $m=l_z=0$; the electron (but not the orbital) is further characterized by a statement of its spin direction, i.e. by specifying $s_z = +\frac{1}{2}$ or $s_z = -\frac{1}{2}$. Two electrons can together occupy the $1s$

orbital provided, according to Rule 1, that one has the set of values $n=1$, $l=0$, $l_z=0$, $s_z=+\frac{1}{2}$, and the other $n=1$, $l=0$, $l_z=0$, $s_z=-\frac{1}{2}$. We talk, rather loosely, of two electrons occupying the same orbital only if their spins are *paired* (or *opposed*). A third electron cannot exist in the same orbital without repeating a set of values for n, l, l_z and s_z already taken up. It would have to be placed into some other orbital and the choice is determined by Rule 2: it would go into the next higher vacant or half-vacant orbital. In general, orbital energies in many-electron atoms increase with increasing n, as they do for hydrogen, but they also increase with increasing l, whereas we noted for hydrogen that all s, p, d,... orbitals with the same n were degenerate. In fact the order of the energy levels for most atoms is as follows:

$$1s < 2s < 2p < 3s < 3p < 4s < 3d < 4p < 5s < 4d \dots \quad (5.16)$$

Thus when the $1s$ orbital is full (i.e. contains two electrons) the next available orbital is the $2s$, and after this the $2p$. Now we remember that there are *three* $2p$ orbitals, one along each co-ordinate axis, and *each* of these can contain two electrons. We may write the n, l, l_z and s_z values as:

$$
\left.
\begin{aligned}
n &= 2, & l &= 1, & l_z &= 1, & s_z &= \pm\tfrac{1}{2} \\
n &= 2, & l &= 1, & l_z &= 0, & s_z &= \pm\tfrac{1}{2} \\
n &= 2, & l &= 1, & l_z &= -1, & s_z &= \pm\tfrac{1}{2}
\end{aligned}
\right\} \text{total six electrons.}
$$

All three p orbitals remain degenerate (as do the five d orbitals, seven f, etc.) for a given n. It is Rule 3 which tells us how electrons occupy these degenerate orbitals. Hund's rule states that when, for example, the $2p_x$ orbital contains an electron, the next electron will go into a *different* $2p$, say $2p_y$, orbital, and a third into the $2p_z$. This may be looked upon as a consequence of repulsion between electrons. A fourth electron has no choice but to pair its spin with an electron already in one $2p$ orbital, while a fifth and sixth will complete the filling of the three $2p$'s.

On this basis we can build up the *electronic configurations* of the ten smallest atoms, from hydrogen to neon. This is shown in Table 5.2 where each box represents an orbital occupied by one or two electrons with spin directions shown by the arrows. A convenient notation for the electronic configuration is also shown in the Table.

When a set of orbitals of given n and l are filled they are referred to as a *closed shell*. Thus the $1s^2$ set of helium, the $2s^2$ set of beryllium and the $2p^6$ set of neon are all closed shells. The convenience of this is that closed shells make no contribution to the orbital or spin angular momentum of the whole atom and hence they may be ignored when discussing atomic spectra. This represents a considerable simplification.

TABLE 5.2: Electronic Structure of some Atoms

	$1s$	$2s$	$2p$			
Hydrogen	↑					$1s^1$
Helium	↑↓					$1s^2$
Lithium	↑↓	↑				$1s^2 2s^1$
Beryllium	↑↓	↑↓				$1s^2 2s^2$
Boron	↑↓	↑↓	↑			$1s^2 2s^2 2p_x^1$
Carbon	↑↓	↑↓	↑	↑		$1s^2 2s^2 2p_x^1 2p_y^1$
Nitrogen	↑↓	↑↓	↑	↑	↑	$1s^2 2s^2 2p_x^1 2p_y^1 2p_z^1$
Oxygen	↑↓	↑↓	↑↓	↑	↑	$1s^2 2s^2 2p_x^2 2p_y^1 2p_z^1$
Fluorine	↑↓	↑↓	↑↓	↑↓	↑	$1s^2 2s^2 2p_x^2 2p_y^2 2p_z^1$
Neon	↑↓	↑↓	↑↓	↑↓	↑↓	$1s^2 2s^2 2p_x^2 2p_y^2 2p_z^2$

3.2. The Spectrum of Lithium and Other Hydrogen-like Species. The alkali metals, lithium, sodium, potassium, rubidium and cesium, all have a single electron outside a closed-shell core (cf. lithium in Table 5.2 above). Superficially, then, they resemble hydrogen and this resemblance is augmented by the fact that we can ignore the angular momentum of the core and deal merely with the spin and orbital momentum of the outer electron. Thus we immediately expect the p, d,... levels to be split into doublets because of coupling between l and s while the s-levels, although technically also doublet levels, are degenerate.

The energy levels of lithium are sketched in Fig. 5.8, which figure should be compared with the corresponding Fig. 5.6 for hydrogen. The two diagrams are similar except for the energy difference

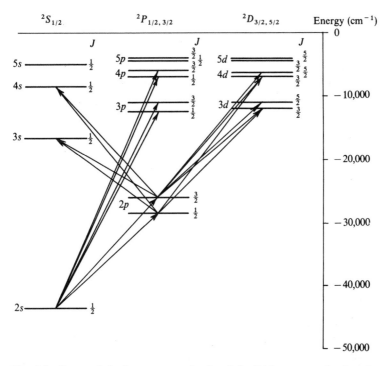

Fig. 5.8: *Some of the lower energy levels of the lithium atom showing the difference in energy of s, p and d states with the same value of n. Some allowed transitions are also shown. The j-splitting is greatly exaggerated.*

between the s, p and d orbitals of given n in the case of lithium, and the fact that, for this metal, the $1s$ state is filled with electrons which do not generally take part in spectroscopic transitions, it requiring much less energy to induce the $2s$ electron to undergo a transition. Under high energy conditions, however, one or both of the $1s$ electrons may be promoted.

The selection rules for alkali metals are the same as for hydrogen, i.e. $\Delta n = \pm 1, \pm 2, \ldots, \Delta l = \pm 1, \Delta j = 0, \pm 1$, and so the spectra will be similar also. Thus transitions from the ground state $(1s^2 2s)$ can occur to p levels: $2S_{1/2} \rightarrow nP_{1/2, 3/2}$, and a series of doublets similar to the Lyman series will be formed converging to some point from which the ionization potential can be found. From the $2p$ state, however, two separate series of lines will be seen:

$$2\,^2P_{1/2, 3/2} \rightarrow n\,^2S_{1/2}$$

and

$$2\,^2P_{1/2, 3/2} \rightarrow n\,^2D_{3/2, 5/2}.$$

The former will be doublets, the latter compound doublets, but their frequencies will differ because the s and d orbital energies are no longer the same.

The same remarks apply to the other alkali metals, the differences between their spectra and that of lithium being a matter of scale only. For instance the j-splitting due to coupling between $\mathbf{l}$ and $\mathbf{s}$ increases markedly with the atomic number. Thus the doublet separation of lines in the spectral series, which is scarcely observable for hydrogen, is less than 1 cm^{-1} for the $2p$ level of lithium, about 17 cm^{-1} for sodium and over 5000 cm^{-1} for cesium.

Any atom which has a single electron moving outside a closed shell will exhibit a spectrum of the type discussed above. Thus ions of the type He$^+$, Be$^+$, B^{2+}, etc., should, and indeed do, show what are termed "hydrogen-like spectra".

4. The Angular Momentum of Many-Electron Atoms

We turn now to consider the contribution of two or more electrons in the outer shell to the total angular momentum of the atom. There are two different ways in which we might sum the orbital and spin momentum of several electrons:

(i) First sum the orbital contributions, and then the spin contributions separately, finally add the total orbital and total spin contributions to reach the grand total. Symbolically:

$$\sum \mathbf{l}_i = \mathbf{L}, \qquad \sum \mathbf{s}_i = \mathbf{S}, \qquad \mathbf{L} + \mathbf{S} = \mathbf{J}$$

where we use bold-face capital letters to designate total momentum.

(*ii*) Sum the orbital and spin momenta of each electron separately, finally summing the individual totals to form the grand total:

$$\mathbf{l}_i + \mathbf{s}_i = \mathbf{j}_i; \qquad \sum \mathbf{j}_i = \mathbf{J}.$$

The first method, known as Russell–Saunders coupling, gives results in accordance with the spectra of small and medium-sized atoms, while the second (called j, j coupling, since individual j's are summed) applies better to large atoms. We shall consider only the former in detail.

4.1. Summation of Orbital Contributions. The orbital momenta, $\mathbf{l}_1, \mathbf{l}_2, \ldots$ of several electrons may be added by the same methods as were discussed in Section 2.3 for the summation of the orbital and spin momenta of a single electron. Thus we could:

(*i*) Add the vectors $\mathbf{l}_1, \mathbf{l}_2, \ldots$ graphically, remembering that their resultant $\mathbf{L}$ must be expressible by

$$\mathbf{L} = \sqrt{L(L+1)} \quad (L = 0, 1, 2, \ldots), \tag{5.17}$$

where L is the total orbital momentum quantum number. Thus $\mathbf{L}$ can have values $0, \sqrt{2}, \sqrt{6}, \sqrt{12}, \ldots$ only. Figure 5.9 illustrates the method for a p and a d electron, $l_1 = 1$, $l_2 = 2$, hence $\mathbf{l}_1 = \sqrt{2}$, $\mathbf{l}_2 = \sqrt{6}$. There are three, and only three, ways in which the two vectors may be combined to give $\mathbf{L}$ consistent with equation (5.17). The three values of $\mathbf{L}$ are seen to be $\sqrt{12}, \sqrt{6}$ and $\sqrt{2}$, corresponding to the quantum number $L = 3, 2$ and 1 respectively.

(*ii*) Alternatively we could add the individual *quantum numbers* l_1 and l_2 to obtain the total quantum number L according to:

$$L = l_1 + l_2, \qquad l_1 + l_2 - 1, \ldots, \qquad |l_1 - l_2|, \tag{5.18}$$

where the modulus sign $|\ldots|$ indicates that we are to take $l_1 - l_2$ or $l_2 - l_1$, whichever is positive. For two electrons, there will plainly be $2l_i + 1$ different values of L, where l_i is the smaller of the two l values.

(*iii*) Finally we could add the z-components of the individual vectors, picking out from the result sets of components corresponding to the various allowed $\mathbf{L}$ values. Symbolically this process is:

$$L_z = \sum l_{i_z}.$$

12

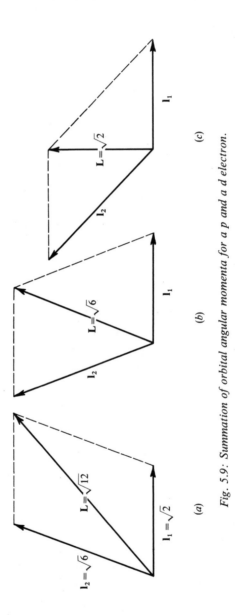

Fig. 5.9: Summation of orbital angular momenta for a p and a d electron.

Of these methods, (*ii*) is the simplest but it is only applicable when the individual electrons concerned have different n or different l values (these are termed *non-equivalent* electrons). If n and l are the same for two or more electrons they are termed *equivalent* and method (iii) must be used. Examples will be given later.

4.2. Summation of Spin Contributions. The same methods may be used here as in the previous sub-section. Briefly, if we write the total spin angular momentum as **S**, and the total spin quantum number as S (which is often simply called the total spin), we can have:

(*i*) Graphical summation, provided the resultant is

$$\mathbf{S} = \sqrt{S(S+1)}, \tag{5.19}$$

where S is either *integral or zero* only, if the number of contributing spins is *even*, or *half-integral* only, if the number is *odd*.

(*ii*) Summation of individual quantum numbers; for N spins we have:

$$S = \sum s_i, \sum s_i - 1, \sum s_i - 2, \dots$$

$$= \frac{N}{2}, \frac{N}{2} - 1, \dots, \frac{1}{2} \quad \text{(for N odd)} \tag{5.20}$$

$$= \frac{N}{2}, \frac{N}{2} - 1, \dots, 0 \quad \text{(for N even)}.$$

(*iii*) Summation of individual s_z to give S_z.

Method (ii), which is *always* applicable, is the simplest. Thus for two electrons we have the two possible spin states:

$$S = \tfrac{1}{2} + \tfrac{1}{2} = 1, \quad \text{or} \quad S = \tfrac{1}{2} + \tfrac{1}{2} - 1 = 0.$$

In the former the spins are called *parallel* and the state may be written (↑↑), while in the latter they are *paired* or *opposed* and written (↑↓).

Again, for three electrons we may have:

$$S = \tfrac{1}{2} + \tfrac{1}{2} + \tfrac{1}{2} = \tfrac{3}{2} \quad (\uparrow\uparrow\uparrow)$$

$$S = \tfrac{1}{2} + \tfrac{1}{2} + \tfrac{1}{2} - 1 = \tfrac{1}{2} \quad (\uparrow\uparrow\downarrow, \uparrow\downarrow\uparrow \text{ or } \downarrow\uparrow\uparrow)$$

where we see that there are three ways in which the $S = \tfrac{1}{2}$ state may be realized, one in which $S = \tfrac{3}{2}$.

As we have implied above, both **L** and **S** have z-components along a reference direction.   For **L** these components are limited to integral values by quantum laws and, as we can see from Fig. 5.10(*a*), there are, in general, $2L+1$ of them, while for **S** the S_z will be integral only or half-integral only, depending on whether S is integral or half-integral.   We show examples in Fig. 5.10(*b*) and (*c*).   In both cases there are $2S+1$ components.

4.3. Total Angular Momentum.   The addition of the total orbital momentum **L** and the total spin momentum **S** to give the grand total momentum **J** can be carried out in the same ways as the addition of **l** and **s** to give **j**, for a single electron.   The only additional point is that the quantum number J in the expression

$$\mathbf{J} = \sqrt{J(J+1)} \tag{5.21}$$

must be *integral* if S is integral, and *half-integral* if S is half-integral. In terms of the quantum numbers we can write immediately:

$$J = L+S, \, L+S-1, \ldots, |L-S| \tag{5.22}$$

where, as before, the *positive* value of $L-S$ is the lowest limit of the series of values.

For example, if $L=2$, $S=\frac{3}{2}$, we would have

$$J = \tfrac{7}{2}, \tfrac{5}{2}, \tfrac{3}{2} \text{ and } \tfrac{1}{2}$$

while if $L=2$, $S=1$, the J values are:

$$J = 3, 2 \text{ or } 1 \text{ only.}$$

In general we see that there are $2S+1$ different values of J and hence $2S+1$ states with different total momentum.   The energy of a state depends on its total momentum, so we arrive at $2S+1$ different energy levels, the energy of each depending on the way in which **L** and **S** are combined.   The quantity $2S+1$, which occupies a special place in atomic spectroscopy, is called the *multiplicity* of the system. We recall that, when discussing the spectrum of atomic hydrogen (Section 2.4) we found each state, other than $l=0$, to be a doublet: in other words, its *multiplicity* is two, a value consistent with $2S+1$ where $S=\frac{1}{2}$ for a single electron.

The $l=0$ state of atomic hydrogen was an apparent exception, but we described it, also, as a doublet state with degenerate energy levels.   In just the same way, if, for several electrons, $L<S$, then there are only $2L+1$ different states.   For example, $L=1$, $S=2$:

$$J = 3, 2 \text{ or } 1 \text{ only,} \qquad (2S+1 = 5, 2L+1 = 3).$$

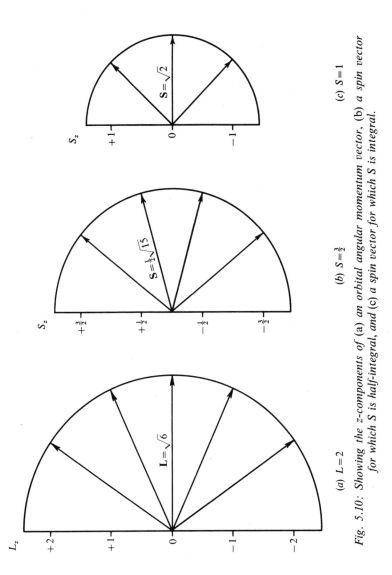

(a) $L = 2$ (b) $S = \frac{3}{2}$ (c) $S = 1$

Fig. 5.10: Showing the z-components of (a) an orbital angular momentum vector, (b) a spin vector for which S is half-integral, and (c) a spin vector for which S is integral.

Even in these cases, however, the multiplicity of the system is still described formally as $2S+1$ with the reservation that some states are degenerate.

4.4. Term Symbols. In the whole of this section we have been describing the way in which the total angular momentum of an atom is built up from its various components. Using one sort of coupling only (the Russell–Saunders coupling) we arrive at vector quantities **L**, **S** and **J** for a system which may be expressed in terms of quantum numbers L, S and J:

$$\mathbf{L} = \sqrt{L(L+1)}; \qquad \mathbf{S} = \sqrt{S(S+1)}; \qquad \mathbf{J} = \sqrt{J(J+1)} \quad (5.23)$$

where the integral L and integral or half-integral S and J are themselves combinations of individual electronic quantum numbers.

In any particular atom, then, we see that the individual electronic angular momenta may be combined in various ways to give different states each having a different total angular momentum (**J**) and hence a different energy (unless some states happen to be degenerate). Before discussing the effect of these states on the spectrum of an atom we require some symbolism which we may use to describe states conveniently. We have already introduced such *state symbols* or *term symbols* in Section 2.6, but we now consider them rather more fully.

The term symbol for a particular atomic state is written as follows:

$$\text{Term symbol} = {}^{2S+1}L_J \qquad (5.24)$$

where the numerical superscript gives the *multiplicity* of the state, the numerical subscript gives the total angular momentum quantum number J, and the value of the orbital quantum number L, is expressed by a *letter*:

$$\text{for } L = 0, 1, 2, 3, 4, \ldots$$

$$\text{symbol} = S, P, D, F, G, \ldots$$

which symbolism is comparable with the s, p, d, ... already used for single electron states with $l=0, 1, 2, \ldots$.

Let us now see some examples.

(i) $S=\frac{1}{2}$, $L=2$; hence $J=\frac{5}{2}$ or $\frac{3}{2}$ and $2S+1=2$. Term symbols: ${}^2D_{5/2}$ and ${}^2D_{3/2}$, which are to be read "doublet D five-halves" and "doublet D three-halves" respectively.

(ii) $S=1$, $L=1$; hence $J=2$, 1 or 0, and $2S+1=3$. Term symbols: $^3P_2$, $^3P_1$ or $^3P_0$ (read "triplet P two", etc.).

In both these examples we see that (since $L \geqslant S$), the multiplicity is the same as the number of different energy states.

(iii) $S=\frac{3}{2}$, $L=1$; hence $J=\frac{5}{2}$, $\frac{3}{2}$ or $\frac{1}{2}$ and $2S+1=4$. Term symbols: $^4P_{5/2}$, $^4P_{3/2}$, $^4P_{1/2}$ (read "quartet P five-halves", etc.) where, since $L<S$, there are only three different energy states but each is nonetheless described as *quartet* since $2S+1=4$.

The reverse process is equally easy; given a term symbol for a particular atomic state we can immediately deduce the various total angular momenta of that state. Some examples:

(iv) $^3S_1$: we read immediately that $2S+1=3$, hence $S=1$, and that $L=0$ and $J=1$.

(v) $^2P_{3/2}$: $L=1$, $J=\frac{3}{2}$, $2S+1=2$, hence $S=\frac{1}{2}$.

Note, however, that the term symbol tells us only the total spin, total orbital and grand total momenta of the whole atom—it tells us nothing of the states of the individual electrons in the atoms, nor even how many electrons contribute to the total. Thus in example (v) above, the fact that $S=\frac{1}{2}$ implies that the atom has an odd number of contributing electrons, all except one of which have their spins paired. Thus a single electron (↑), three electrons (↑↑↓), five electrons (↑↓↑↓↑) etc., all form a doublet state. Similarly, the value $L=1$ implies, perhaps, one p electron, or perhaps one p and two s electrons, or one of many other possible combinations.

Normally this is not important; the spectroscopist is interested only in the energy state of the atom as a whole. Should we wish to specify the energy states of individual electrons, however, we can do so by including them in the term symbol as a prefix. Thus in example (v) we might have $2p\ ^2P_{3/2}$, or $1s2p3s\ ^2P_{3/2}$, etc.

We can now apply our knowledge of atomic states to the discussion of the spectra of some atoms with two or more electrons. We start with the simplest, that of helium.

4.5. The Spectrum of Helium and the Alkaline Earths. Helium, atomic number two, consists of a central nucleus and two outer electrons. If we place one of these electrons into the lowest energy orbital (the $1s$), Pauli's principle tells us that the second electron can do one of two things.

(i) If the spins of the two electrons are *opposed* ($s_{1_z} = +\frac{1}{2}$, $s_{2_z} = -\frac{1}{2}$, say) then they can *both* occupy the 1s orbital. We could designate them, then, as:

electron 1: $n_1 = 1$, $l_1 = 0$, $l_{1_z}(\equiv m_1) = 0$, $s_{1_z} = +\frac{1}{2}$
electron 2: $n_2 = 1$, $l_2 = 0$, $l_{2_z}(\equiv m_2) = 0$, $s_{2_z} = -\frac{1}{2}$.

(ii) If their spins are *parallel* ($s_{1_z} = +\frac{1}{2} = s_{2_z}$) then they cannot both occupy the same orbital and, while electron 1 is in the 1s orbital: $n_1 = 1$, $l_1 = 0$, $l_{1_z} = 0$, $s_{1_z} = +\frac{1}{2}$, electron 2 must have $n_2 > 1$ and it can then have a large number of different values for l_2 and l_{2_z}.

In (i) both electrons occupy the orbital of lowest energy whereas in (ii) one of them is in a higher energy state. Plainly (i) is the *ground state* of the atom, and we shall consider this first. In the ground state we have $s_{1_z} + s_{2_z} = 0$, hence $S_z = 0$ and $S = 0$. Thus the ground state is a *singlet* state ($2S + 1 = 1$). Further $L = l_1 + l_2 = 0$ and hence J can only be zero. Thus the ground state is $^1S_0$ for helium.

Let us, for the moment, allow one electron only to undergo transitions to higher orbitals; we sketch the energy levels of this electron on the left-hand side of Fig. 5.11, together with some of the allowed transitions. The selection rules in general for such transitions are:

$$\Delta S = 0, \quad \Delta L = \pm 1, \quad \Delta J = 0, \pm 1. \qquad (5.25)$$

(There is a further rule that a state with $J = 0$ cannot make a transition to another $J = 0$ state, but this will not concern us here.) We see immediately that, since S cannot change during a transition, the singlet ground state can undergo transitions only to other singlet states. The selection rules for L and J are the same as those for l and j considered earlier.

Initially the $1s^2$ $^1S_0$ state can undergo a transition only to $1s^1np^1$ states (since we are allowing only one electron to change its orbital). In the $1s^1np^1$ states we have, plainly, $L = 1$, $S = 0$, hence $J = 1$ only. The complete transition may thus be symbolized:

$$1s^2 \, ^1S_0 \rightarrow 1s^1np^1 \, ^1P_1$$

or, briefly:

$$^1S_0 \rightarrow \, ^1P_1.$$

From the $^1P_1$ state the system could either revert to $^1S_0$ states, as shown in the figure, or undergo transitions to the higher $^1D_2$ states (for these $S=0$, $L=2$, hence $J=2$ only). In general, then, all these transitions will give rise to spectral series very similar to those of lithium except that here transitions are between *singlet* states only and all the spectral lines will be single.

Let us now return to the situation in which the electron spins are parallel and hence the electrons cannot occupy the same orbital.

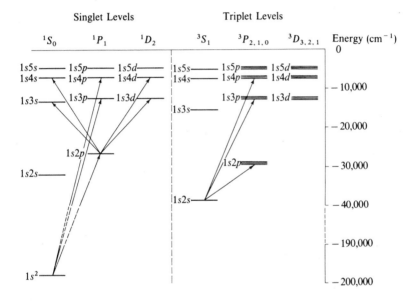

Fig. 5.11: *Some of the energy levels of the electrons in the helium atom together with a few allowed transitions.*

Here we have $s_{1_z}+s_{2_z}=S_z=1$, and hence all the states will be *triplet*. The lowest such state will be with one electron in the $1s$ orbital, the other in the $2s$:

electron 1: $n_1 = 1$, $l_1 = 0$, $l_{1_z} = 0$
electron 2: $n_2 = 2$, $l_2 = 0$, $l_{2_z} = 0$.

Hence we have $L=l_1+l_2=0$, and $J=L+S=1$ only. Thus the state is $^3S_1$. The selection rules of equation (5.25) show that this state can undergo transitions only to another triplet state and that the final state must have $L=1$, and hence be a P state. With $S=1$,

$L = 1$ we can have three values of J: 2, 1 and 0.   Hence the transition is

$$^3S_1 \rightarrow {}^3P_2, \; {}^3P_1 \; \text{or} \; {}^3P_0.$$

All three transitions are allowed, since $\Delta J = 0$ or ± 1, hence the resulting spectral line will be a *triplet*.

We show the situation on the right-hand side of Fig. 5.11 where the $1s2s \; {}^3S_1$ state (remember that one electron is *confined to the 1s orbital* for this discussion) is shown as the lowest energy triplet state. Transitions from this to $1s \, np \; {}^3P_{2,1,0}$ states lead to a spectral series, similar to those for lithium, with each line a triplet.

Transitions from the ^3P states may take place either to ^3S states (spectral series of triplets) or to ^3D states.   In the latter case the spectral series may be very complex if completely resolved.   For ^3D we have $S = 1$, $L = 2$, hence $J = 3$, 2 or 1 and we show in Fig. 5.12 a transition between ^3P and ^3D states, bearing in mind the selection rule $\Delta J = 0, \; \pm 1$.   We note that $^3P_2$ can go to each of $^3D_{3,2,1}$, $^3P_1$ can go only to $^3D_{2,1}$ and $^3P_0$ can go only to $^3D_1$.   Thus the complete spectrum (shown at the foot of the figure) should consist of six lines.   Normally, however, the very close spacing is not resolved, and only three lines are seen; for this reason the spectrum is referred to as a *compound triplet*.

We might note in passing that Fig. 5.12 shows that levels with higher J have a higher energy in helium, and that the separation decreases from top to bottom.   This is not the case with all atoms, however.   If higher J is equivalent to lower energy then the separation *increases* from top to bottom and the multiplet is described as *inverted*.   In helium, and other atoms with similar behaviour, the multiplet is *normal* or *regular*.

We see, then, that the spectrum of helium consists of spectral series grouped into two types which overlap each other in frequency. In one type, involving transitions between singlet levels, all the spectral lines are themselves singlets, while in the other the transitions are between triplet states and each "line" is at least a close triplet and possibly even more complex.   Because of the selection rule $\Delta S = 0$ there is a strong prohibition on transitions between singlet and triplet states and transitions *cannot occur* between the right- and left-hand sides of Fig. 5.11.   Early experimenters, noting the difference between the two types of spectral series, suggested that helium exists in two modifications, ortho- and para-helium. This is not far from the truth, although we know now that the difference between the two forms is very subtle: it is merely that one has its electron spins always opposed, and the other always parallel.

Other atoms containing two outer electrons exhibit spectra similar to that of helium. Thus the alkaline earths, beryllium, magnesium, calcium, etc., fall in this category, as do ionized species with just two remaining electrons, e.g. B^+, C^{2+}, etc.

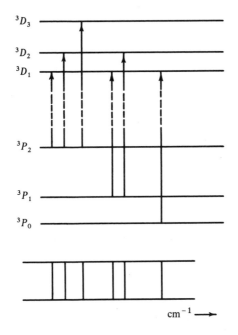

Fig. 5.12: The "compound triplet" spectrum arising from transitions between ^3P and ^3D levels in the helium atom. The separation between levels of different J is much exaggerated.

We should remind the reader at this point that the above discussion on helium has been carried through on the assumption that one electron remains in the 1s orbital all the time. This is a reasonable assumption since a great deal of energy would be required to excite two electrons simultaneously, and this would not happen under normal spectroscopic conditions. However, not all atoms have only s electrons in their ground state configuration, and we consider next some of the consequences.

4.6. Equivalent and Non-Equivalent Electrons; Energy Levels of Carbon. The ground state electronic configuration of carbon is $1s^2 2s^2 2p^2$, which indicates that both the $1s$ and $2s$ orbitals are filled (and hence contribute nothing to the angular momentum of the atom) while the $2p$ orbitals are only partially filled. The $2p$ electrons, also, are most easily removed, so it is these which normally undergo spectroscopic transitions.

Two or more electrons are referred to as *equivalent* if they have the same value of n and of l. Thus the two $2p$ electrons in the ground state of carbon are equivalent ($n_1 = n_2 = 2$, $l_1 = l_2 = 1$) while the set of $1s2s$ are non-equivalent ($n_1 \neq n_2$ although $l_1 = l_2$) as are, for example, $2s2p$ ($n_1 = n_2$ but $l_1 \neq l_2$). Special care is necessary when considering the total angular momentum of equivalent electrons since restrictions are placed on the values of the quantum numbers which each may have. Let us consider the case of $2p^2$ in some detail.

The first restriction arises from Pauli's principle (Section 3.1). Since we have $n_1 = n_2$ and $l_1 = l_2$ we cannot simultaneously choose $l_{1_z} = l_{2_z}$ *and* $s_{1_z} = s_{2_z}$.

Further restrictions follow from physical considerations. The basic principle is that electrons cannot be distinguished from each other and so if the energies of two electrons are exchanged we have no way of discovering experimentally that such exchange has taken place. The implication is that if the values of all four numbers n, l, l_z and s_z for each of two electrons are exchanged, the initial situation is identical in every way with the final. When considering total momentum and the term symbols of atoms we are interested only in different situations and we must not count twice those systems which are interconvertible merely by an exchange of all four numbers n, l, l_z and s_z. Consider some examples chosen from the $2p^2$ case:

(i) We have $n_1 = n_2$, $l_1 = l_2$ and if we also choose $l_{1_z} = l_{2_z}$ then we know (Pauli's principle) that if $s_{1_z} = +\frac{1}{2}$, then $s_{2_z} = -\frac{1}{2}$; alternatively if $s_{1_z} = -\frac{1}{2}$ then $s_{2_z} = +\frac{1}{2}$. Now these two cases are completely identical because one can be reached from the other by exchanging n, l, l_z and s_z. Thus, while we can consider either set alone as typical of the state, we must not consider both together.

(ii) Similarly, if we assume $s_{1_z} = s_{2_z}$ then we know $l_{1_z} \neq l_{2_z}$. For

p electrons $l=1$ and hence $l_z = 1, 0$ or -1. So we might have:

$$s_{1_z} = s_{2_z}: \quad l_{1_z} = 1; \quad l_{2_z} = 0 \text{ or } -1$$
$$l_{1_z} = 0; \quad l_{2_z} = 1 \text{ or } -1$$
$$l_{1_z} = -1; \quad l_{2_z} = 1 \text{ or } 0.$$

Note, however, that the system represented by the pair of values $(1, 0)$ for (l_{1_z}, l_{2_z}) is identical with that for $(0, 1)$; $(0, -1)$ is identical with $(-1, 0)$ and $(1, -1)$ with $(-1, 1)$. Thus we reduce the above six pairs to only three *different* sets:

$$s_{1_z} = s_{2_z}: \quad l_{1_z} = 1; \quad l_{2_z} = 0$$
$$l_{1_z} = 1; \quad l_{2_z} = -1$$
$$l_{1_z} = 0; \quad l_{2_z} = -1.$$

(*iii*) Finally we note that if $l_{1_z} \neq l_{2_z}$ and $s_{1_z} \neq s_{2_z}$, then interchange of only *one* pair of values (e.g. $s_{1_z} = +\frac{1}{2}$, $s_{2_z} = -\frac{1}{2}$, $\rightarrow s_{1_z} = -\frac{1}{2}$, $s_{2_z} = +\frac{1}{2}$) does produce a *different* situation; physically an electron already distinguishable by its l_z value is being reversed in spin. All four numbers n, l, l_z and s_z must be exchanged to produce an indistinguishable state.

Keeping these rules in mind we can construct Table 5.3, in which the first four columns list combinations of l_{1_z}, l_{2_z} and s_{1_z} and s_{2_z} leading to different energy states (let us call them sub-states). We are interested in the *total* energy and so we show in the next two columns the values of $l_{1_z} + l_{2_z} = L_z$ and $s_{1_z} + s_{2_z} = S_z$ respectively. The final column merely supplies a convenient label to each sub-state for the following discussion. The fifteen sub-states in the table constitute the z-components of the various $\mathbf{L}$ and $\mathbf{S}$ vectors which may be formed from two equivalent p electrons. We can find the term symbols in the following way:

(*i*) Note first that the largest L_z value in the table is $L_z = +2$ (sub-state (*a*)), and this is associated with $S_z = 0$. $L_z = +2$ must be a z-component of the state $L=2$, i.e. one component of a D state, the other components of which are $L_z = +1, 0, -1$ and -2. In the table we can find several sub-states of the requisite L_z values, all associated with $S_z = 0$; it is immaterial which of the alternatives we choose, so let us take sub-states (*c*), (*g*), (*l*) and (*o*). Since $S_z = 0$ the state must be singlet, hence we have:

$$^1D = \text{sub-states } (a), (c), (g), (l) \text{ and } (o).$$

With $L=2$, $S=0$ this can only be a $^1D_2$ state.

(ii) Of the remaining sub-states the largest L_z is $+1$ associated with $S_z = +1$ (sub-state (b)). This is plainly one component of a ^3P state, the other components of which may be selected as:

$$^3P \text{ state:} \quad S_z = +1: \quad L_z = 1, 0 \text{ and } -1, \text{ i.e. } (b), (f), (k)$$

$$S_z = 0: \quad L_z = 1, 0 \text{ and } -1, \text{ i.e. } (d), (h), (m)$$

$$S_z = -1: \quad L_z = 1, 0 \text{ and } -1, \text{ i.e. } (e), (j), (n).$$

Here we have considered the three components of $S=1$, $S_z = +1$, 0 and -1, to be associated in turn with the components of $L=1$. The three states listed correspond to term symbols $^3P_2$, $^3P_1$ and $^3P_0$.

TABLE 5.3: Sub-States of Two Equivalent p Electrons
$(n_1 = n_2 = 2; l_1 = l_2 = 1)$

l_{1_z}	l_{2_z}	s_{1_z}	s_{2_z}	L_z $(l_{1_z} + l_{2_z})$	S_z $(s_{1_z} + s_{2_z})$	Sub-State
$+1$	$+1$	$+\frac{1}{2}$	$-\frac{1}{2}$	$+2$	0	(a)
$+1$	0	$+\frac{1}{2}$	$+\frac{1}{2}$	$+1$	$+1$	(b)
$+1$	0	$+\frac{1}{2}$	$-\frac{1}{2}$	$+1$	0	(c)
$+1$	0	$-\frac{1}{2}$	$+\frac{1}{2}$	$+1$	0	(d)
$+1$	0	$-\frac{1}{2}$	$-\frac{1}{2}$	$+1$	-1	(e)
$+1$	-1	$+\frac{1}{2}$	$+\frac{1}{2}$	0	$+1$	(f)
$+1$	-1	$+\frac{1}{2}$	$-\frac{1}{2}$	0	0	(g)
$+1$	-1	$-\frac{1}{2}$	$+\frac{1}{2}$	0	0	(h)
$+1$	-1	$-\frac{1}{2}$	$-\frac{1}{2}$	0	-1	(i)
0	0	$+\frac{1}{2}$	$-\frac{1}{2}$	0	0	(j)
0	-1	$+\frac{1}{2}$	$+\frac{1}{2}$	-1	$+1$	(k)
0	-1	$+\frac{1}{2}$	$-\frac{1}{2}$	-1	0	(l)
0	-1	$-\frac{1}{2}$	$+\frac{1}{2}$	-1	0	(m)
0	-1	$-\frac{1}{2}$	$-\frac{1}{2}$	-1	-1	(n)
-1	-1	$+\frac{1}{2}$	$-\frac{1}{2}$	-2	0	(o)

(iii) Finally the one remaining sub-state, (j), has $L_z = S_z = 0$ and this plainly comprises a $^1S_0$ state.

Overall, then, two equivalent p electrons give rise to the three different energy states ^1D, ^1S and ^3P, of which the latter, being a triplet state, has three close energy levels $^3P_2$, $^3P_1$ and $^3P_0$. Hund's rule, which we quoted in Section 3.1 may be expressed for equivalent electrons as: "the state of lowest energy for a given electronic configuration is that having the greatest multiplicity. If more than one state has the same multiplicity then the lowest of these is that with the greatest L value ".

Thus for carbon the ground state is the ^3P, the next in energy is the ^1D and finally the ^1S. We note that this new expression of Hund's rule implies that electrons in degenerate orbitals tend to have their spins parallel (since this gives the greatest multiplicity, and hence lowest energy); this in turn means that electrons tend to go into separate orbitals since in the same orbital they must have paired spins. Thus we are justified in writing the electronic structures of carbon, nitrogen and oxygen as in Table 5.2.

If now one of the $2p$ electrons of carbon is promoted to the $3p$ state we have the configuration $1s^2 2s^2 2p3p$. This is an excited state in which the p electrons are non-equivalent. The interested student should show, by the method of Table 5.3, that six different term symbols can be found for this configuration, i.e. ^1S, ^1P, ^1D, ^3S, ^3P and ^3D. In this case, since $n_1 \neq n_2$, neither the Pauli principle nor the principle of indistinguishability offers restrictions to l_z and s_z values and hence more terms result.

For non-equivalent electrons, however, it is simpler to deal directly with L and S values. Thus we have $s_1 + s_2 = 1$ or 0 depending upon whether the electron spins are parallel or opposed, while for $l_1 = l_2 = 1$ we can have $L = 2, 1$ or 0. We can then tabulate L, S and J directly, together with their term symbols:

L	S	J	Term symbol
2	1	3, 2 or 1	$^3D_{3,2,1}$
2	0	2	$^1D_2$
1	1	2, 1 or 0	$^3P_{2,1,0}$
1	0	1	$^1P_1$
0	1	1	$^3S_1$ (triply degenerate)
0	0	0	$^1S_0$

and we arrive at the six states listed previously. Note that this direct method is not applicable to equivalent electrons because summation of l to give L implies that all l_z are allowed: this, we have seen, is not true when the electrons are equivalent.

Many other electronic configurations occur, both for carbon and other atoms, in which two or more equivalent electrons contribute to the total energy. We shall not discuss these further, however, except to state that their total energies and term symbols may be discovered by the same process as exemplified above for $2p^2$ electrons.

5. The Zeeman Effect

We have been concerned in this chapter with two sorts of electronic energy. Firstly, there is energy of *position*—energy arising by virtue of interaction between electrons and the nucleus and between electrons and other electrons in the same atom. This energy can be described in terms of the n and l quantum numbers, although we have discussed it, rather less precisely, by drawing energy level diagrams. Secondly, there is energy of *motion*—energy arising from the summed orbital and spin momenta of the electrons in the atom which depends on the l_z and s_z values of each electron and the way in which these are coupled. This gives rise to the fine structure of spectroscopic lines discussed earlier.

Angular momentum can be considered as arising from a physical movement of electrons about the nucleus and, since electrons are charged, such motion constitutes a circulating electric current and hence a magnetic field. This field can, indeed, be detected, and it is its interaction with exterior fields which is the subject of this section.

We can represent the angular momentum field by a vector $\boldsymbol{\mu}$—the magnetic dipole of the atom—and it is readily shown that $\boldsymbol{\mu}$ is directly proportional to the angular momentum $\mathbf{J}$ and has the same direction. If the electron is considered as a point of mass m and charge e, then we have:

$$\boldsymbol{\mu} = -\frac{e}{2mc}\mathbf{J} \text{ erg. gauss}^{-1}. \qquad (5.26)$$

But quantum mechanics indicates that the electron is not a point charge and a more exact expression for $\boldsymbol{\mu}$ is:

$$\boldsymbol{\mu} = -\frac{ge}{2mc}\mathbf{J} = -\frac{ge}{2mc}\sqrt{J(J+1)}\frac{h}{2\pi} \text{ erg. gauss}^{-1}, \qquad (5.27)$$

where g is a purely numerical factor, called the Landé splitting factor. This factor depends on the state of the electrons in the atom and is given by:

$$g = \frac{3}{2}+\frac{S(S+1)-L(L+1)}{2J(J+1)}. \qquad (5.28)$$

In general g lies between 0 and 2.

We now recall (cf. equation (5.12) for one electron) that $\mathbf{J}$ can have either integral or half-integral components J_z along a reference direction, depending upon whether the quantum number J is integral or half-integral. Figure 5.13(a) shows this for a state with $J = \frac{3}{2}$, the $2J+1$ components being given in general by:

$$J_z = J, J-1, \ldots, \tfrac{1}{2} \text{ or } 0, \ldots, -J. \tag{5.29}$$

Further, since $\mathbf{\mu}$ is proportional to $\mathbf{J}$, $\mathbf{\mu}$ will also have components in the z-direction which are given by:

$$\mu_z = -\frac{ge}{2mc}\frac{h}{2\pi}J_z. \tag{5.30}$$

These are shown diagrammatically at Fig. 5.13(b). If now an external field is applied to the atom, thus specifying the previously arbitrary z-direction, the atomic dipole $\mathbf{\mu}$ will interact with the applied field to an extent depending on its component in the field direction. If the strength of the applied field is H_z then the extent of the interaction is simply $\mu_z H_z$:

$$\text{Interaction} = \Delta E = \mu_z H_z = -\frac{heg}{4\pi mc}H_z \text{ ergs.} \tag{5.31}$$

In this equation we have expressed the interaction as ΔE since the application of the field splits the originally degenerate energy levels corresponding to the $2J+1$ values of J_z into $2J+1$ *different* energy levels. This is shown for $J = \frac{3}{2}$ in Fig. 5.13(c). It is this splitting, or lifting of the degeneracy on the application of an external magnetic field, which is called the *Zeeman effect* after its discoverer.

The energy splitting is very small; the factor $he/4\pi mc$ in equation (5.31), known as the *Bohr magneton*, has a value of 0.927×10^{-20} erg. gauss^{-1}; thus for $g=1$ and $H_z = 10,000$ gauss, the interaction energy ΔE is only some 10^{-17} erg, which in turn is of the order of 0.2 cm^{-1}. This small splitting is, of course, reflected in a splitting of the spectral transitions observed when a magnetic field is applied to an atom. In order to discuss the effect on the spectrum we need one further selection rule:

$$\Delta J_z = 0, \pm 1.$$

Let us consider the doublet lines in the sodium spectrum produced, as we have discussed in Section 3.2, by transitions between the (degenerate) $^2S_{1/2}$ states and the (non-degenerate) $^2P_{1/2}$ and $^2P_{3/2}$ states. When a field H_z is applied to the atom, the $^2S_{1/2}$ and

13

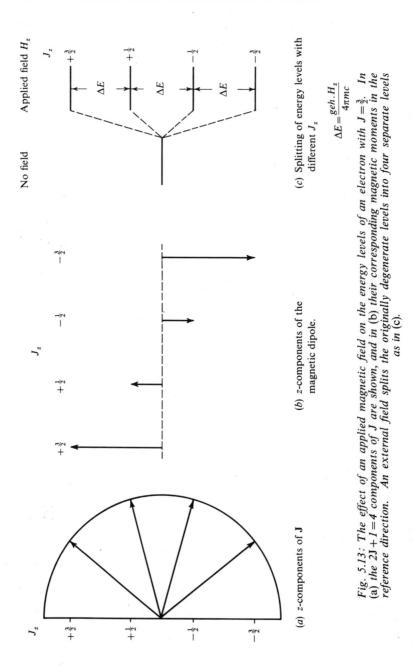

(a) z-components of **J**

(b) z-components of the magnetic dipole.

(c) Splitting of energy levels with different J_z

$$\Delta E = \frac{geh.H_z}{4\pi mc}$$

Fig. 5.13: The effect of an applied magnetic field on the energy levels of an electron with $J = \frac{3}{2}$. In (a) the $2J+1 = 4$ components of J are shown, and in (b) their corresponding magnetic moments in the reference direction. An external field splits the originally degenerate levels into four separate levels as in (c).

$^2P_{1/2}$ states are both split into two (since $J = \frac{1}{2}$, $2J + 1 = 2$), while the $^2P_{3/2}$ is split into four. The extent of the splitting (equation (5.31)) is proportional to the g-factor in each state and, from equation (5.28) we can easily calculate:

$$^2S_{1/2}: \quad S = \tfrac{1}{2},\ L = 0,\ J = \tfrac{1}{2},\ \text{hence}\ g = 2$$

$$^2P_{1/2}: \quad S = \tfrac{1}{2},\ L = 1,\ J = \tfrac{1}{2},\ \text{hence}\ g = \tfrac{2}{3}$$

$$^2P_{3/2}: \quad S = \tfrac{1}{2},\ L = 1,\ J = \tfrac{3}{2},\ \text{hence}\ g = 1\tfrac{1}{3}$$

and we see that the $^2S_{1/2}$, $^2P_{1/2}$ and $^2P_{3/2}$ levels are split in the ratio of $3:1:2$. We show the situation in Fig. 5.14. On the left of the figure we see the energy levels and transitions *before* the field H_z is applied; the levels are unsplit and the spectrum is a simple doublet.

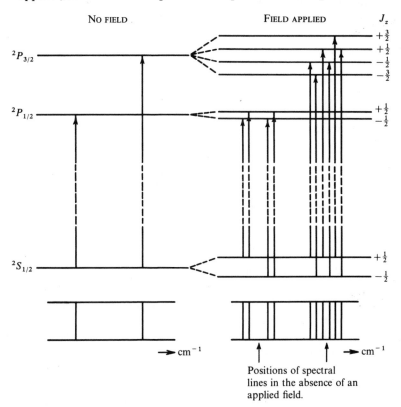

Fig. 5.14: The Zeeman effect on transitions between ^2S and ^2P states. The situation before the field is applied is shown on the left, that after on the right.

On the right we see the effect of the applied field. The spectrum shows that the original line due to the $^2S_{1/2} \rightarrow {}^2P_{1/2}$ transition disappears and is replaced by *four* new lines, while the $^2S_{1/2} \rightarrow {}^2P_{3/2}$ transition is replaced by *six* new lines.

The effect described above is usually referred to as the *anomalous Zeeman effect*—although, in fact, most atoms show the effect in this form. The *normal Zeeman effect* applies to transitions between singlet states only (e.g. the transitions of electrons in the helium atom shown on the left of Fig. 5.11). For singlet states we have:

$$2S + 1 = 1, \quad \text{hence} \quad S = 0.$$
$$\therefore \ J = L \quad \text{and} \quad g = 1 \quad \text{(cf. equation (5.28))}.$$

Thus the splitting between all singlet levels is identical for a given applied field and the corresponding Zeeman spectrum is considerably simplified.

In general, the Zeeman effect can give very useful information about the electronic states of atoms. In the first place, the number of lines into which each transition becomes split when a field is applied depends on the *J*-value of the states between which transitions arise. Next the *g*-value, deduced from the splitting for a known applied field, gives information about the *L* and *S* values of the electron undergoing transitions. Overall, then, the term symbols for various atomic states can be deduced by Zeeman experiments. In this way all the details of atomic states, term symbols, etc., discussed above, have been amply confirmed experimentally.

6. The Influence of Nuclear Spin

The nuclei of many atoms are known to be spinning about an axis. We shall discuss at some length in Chapter 7 the spectrum which this spin may give rise to in the radiofrequency region, but it is pertinent here to consider very briefly what effect such spin may have on the electronic spectra of atoms.

The nuclear spin quantum number *I* may be zero, integral or half integral depending on the particular nucleus considered. Thus the nuclear angular momentum, given by:

$$\mathbf{I} = \sqrt{I(I+1)} \frac{h}{2\pi} = \sqrt{I(I+1)} \text{ a.m. units} \qquad (5.32)$$

can have values 0, $\sqrt{3}/2$, $\sqrt{2}$, $\sqrt{15}/2$, etc.

The effect of I on the spectrum can be understood if we define the total momentum (electronic + nuclear) of an atom by F:

$$F = \sqrt{F(F+1)}\,\frac{h}{2\pi} = \sqrt{F(F+1)} \text{ a.m. units,} \qquad (5.33)$$

where F is the *total momentum quantum number*. If, as before, J is the total electronic quantum number, then we may write

$$F = J+I, J+I-1, \ldots, |J-I| \qquad (5.34)$$

thus giving $2J+1$ or $2I+1$ different energy states, whichever is the less.

The energy level splitting due to nuclear spin is of the order 10^{-3} that due to electron spin; thus extremely fine resolving power is necessary for its observation and it is normally referred to as *hyperfine structure*.

7. Conclusion

This completes all we have to say about atomic spectroscopy.   In the next chapter we extend the ideas introduced here to cover the electronic spectra of simple molecules, and we shall briefly discuss the techniques of electronic spectroscopy.

Bibliography

See the Bibliography to Chapter 6 with the following important additions:

HERZBERG, G.: *Atomic Spectra and Atomic Structure*, Dover, 1944
KUHN, M. G.: *Atomic Spectra*, Academic Press, 1962

CHAPTER 6

ELECTRONIC SPECTROSCOPY
OF MOLECULES

In the first section of this chapter we shall discuss, in some detail, the electronic spectra of diatomic molecules. We shall find that the overall appearance of such spectra can be considered without assuming any knowledge of molecular structure, without reference to any particular electronic transition, and indeed, with little more than a formal understanding of the nature of electronic transitions within molecules. In Section 2 we shall summarize modern ideas of molecular structure and show how these lead to a classification of electronic states analogous to the classification of atomic states discussed in the previous chapter. Section 3 will extend the ideas of Sections 1 and 2 to polyatomic molecules and Section 4 will deal briefly with experimental techniques.

1. Electronic Spectra of Diatomic Molecules

1.1. The Born–Oppenheimer Approximation. As a first approach to the electronic spectra of diatomic molecules we may use the Born–Oppenheimer approximation previously mentioned in Chapter 3, Section 2; in the present context this may be written:

$$E_{\text{total}} = E_{\text{electronic}} + E_{\text{vibration}} + E_{\text{rotation}} \qquad (6.1)$$

which implies that the electronic, vibrational and rotational energies of a molecule are completely independent of each other. We shall see later to what extent this approximation is invalid. A change in the total energy of a molecule may then be written:

$$\Delta E_{\text{total}} = \Delta E_{\text{elec.}} + \Delta E_{\text{vib.}} + \Delta E_{\text{rot.}} \text{ ergs}$$

or

$$\Delta \varepsilon_{\text{total}} = \Delta \varepsilon_{\text{elec.}} + \Delta \varepsilon_{\text{vib.}} + \Delta \varepsilon_{\text{rot.}} \text{ cm}^{-1}. \qquad (6.2)$$

The approximate orders of magnitude of these changes are:

$$\Delta \varepsilon_{\text{elec.}} \approx \Delta \varepsilon_{\text{vib.}} \times 10^3 \approx \Delta \varepsilon_{\text{rot.}} \times 10^6 \text{ cm}^{-1} \qquad (6.3)$$

and so we see that vibrational changes will produce a "coarse structure" and rotational changes a "fine structure" on the spectra of electronic transitions. We should also note that whereas pure rotation spectra (Chapter 2) are shown only by molecules possessing a permanent electric dipole moment, and vibrational spectra (Chapter 3) require a change of dipole during the motion, electronic spectra are given by *all* molecules since changes in the electron distribution in a molecule are always accompanied by a dipole change. This means that homonuclear molecules (e.g. H_2 or N_2), which show no rotation or vibration–rotation spectra, *do* give an electronic spectrum and show vibrational and rotational structure in their spectra from which rotational constants and bond vibration frequencies may be derived.

Initially we shall ignore rotational fine structure and discuss the appearance of the vibrational coarse structure of spectra.

1.2. Vibrational Coarse Structure: Progressions. Ignoring rotational changes means that we rewrite equation (6.1) as

$$E_{\text{total}} = E_{\text{elec.}} + E_{\text{vib.}} \text{ ergs}$$

or

$$\varepsilon_{\text{total}} = \varepsilon_{\text{elec.}} + \varepsilon_{\text{vib.}} \text{ cm}^{-1}. \qquad (6.4)$$

From equation (3.12) we can write immediately:

$$\varepsilon_{\text{total}} = \varepsilon_{\text{elec.}} + (v + \tfrac{1}{2})\bar{\omega}_e - x_e(v + \tfrac{1}{2})^2 \bar{\omega}_e \text{ cm}^{-1} \quad (v = 0, 1, 2, \ldots). \qquad (6.5)$$

The energy levels of this equation are shown in Fig. 6.1 for two arbitrary values of $\varepsilon_{\text{elec.}}$. As in previous chapters the lower states are distinguished by a *double* prime (v'', $\varepsilon''_{\text{elec.}}$), while the upper states carry only a *single* prime (v', $\varepsilon'_{\text{elec.}}$). Note that such a diagram cannot show correctly the relative separations between levels of different $\varepsilon_{\text{elec.}}$, on the one hand, and those with different v' or v'' on the other (cf. equation (6.3)), but that the spacing between the upper vibrational levels is deliberately shown to be rather smaller than that between the lower; this is the normal situation since an excited electronic state usually corresponds to a weaker bond in the molecule and hence a smaller vibrational wavenumber $\bar{\omega}_e$.

There is essentially no selection rule for v when a molecule undergoes an electronic transition, i.e. every transition $v'' \rightarrow v'$ has some probability, and a great many spectral lines would, therefore, be expected. However, the situation is considerably simplified if the *absorption* spectrum is considered from the electronic ground state. In this case, as we have seen in Chapter 3, Section 1.3, virtually all the molecules exist in the lowest vibrational state, i.e. $v'' = 0$, and so the only transitions to be observed with appreciable intensity are those indicated in Fig. 6.1. These are conventionally labelled according to their (v', v'') numbers (note: upper state *first*), i.e. (0, 0), (1, 0), (2, 0), etc. Such a set of transitions is called a *band*

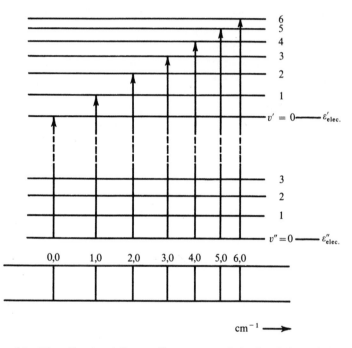

Fig. 6.1: The vibrational "coarse" structure of the band formed during electronic absorption from the ground ($v''=0$) state to a higher state.

since, under low resolution, each line of the set appears somewhat broad and diffuse, and is more particularly called a v' *progression*, since the value of v' increases by unity for each line in the set. The diagram shows that the lines in a band crowd together more closely at high frequencies; this is a direct consequence of the anharmonicity of the upper state vibration which causes the excited vibrational levels to converge.

An analytical expression can easily be written for this spectrum.

From equation (6.5) we have immediately:

$$\Delta\varepsilon_{\text{total}} = \Delta\varepsilon_{\text{elec.}} + \Delta\varepsilon_{\text{vib.}}$$

$$\therefore \; \bar{v}_{\text{spec.}} = (\varepsilon' - \varepsilon'') + \{(v' + \tfrac{1}{2})\bar{\omega}'_e - x'_e(v' + \tfrac{1}{2})^2 \bar{\omega}'_e\}$$

$$- \{(v'' + \tfrac{1}{2})\bar{\omega}''_e - x''_e(v'' + \tfrac{1}{2})^2 \bar{\omega}''_e\} \; \text{cm}^{-1} \qquad (6.6)$$

and, provided some half-dozen lines can be observed in the band, values for $\bar{\omega}'_e$, x'_e, $\bar{\omega}''_e$ and x''_e, as well as the separation between electronic states, $(\varepsilon' - \varepsilon'')$, can be calculated. Thus the observation of a band spectrum leads not only to values of the vibrational frequency and anharmonicity constant in the ground state ($\bar{\omega}''_e$ and x''_e), but also to these parameters in the excited electronic state ($\bar{\omega}'_e$ and x'_e). This latter information is particularly valuable since such excited states may be extremely unstable and the molecule may exist in them for very short times; nonetheless the band spectrum can tell us a great deal about the bond strength of such species.

Of course, most vibrational–electronic spectra are rather more complicated than Fig. 6.1 would suggest. In the first place, some molecules exist in the $v'' = 1$ state initially and, if these undergo absorption, a progression of the type (0, 1), (1, 1), (2, 1), etc., will be formed. This will overlap the progression shown in the figure but, at least in an absorption spectrum, it will be weak and readily distinguished.

In emission, however, the molecule may have been previously excited to any ε', v' level and it can revert to any ε'', v'' state (but see next section for relative transition probabilities), so a great many v' progressions will be observed. Further we shall see later that there are many excited electronic levels, ε', for a molecule, just as there are for atoms, and so in absorption or emission, transitions can take place between several different electronic states; each transition will be accompanied by vibrational structure, and so a particular molecule will give rise to many bands like that shown in Fig. 6.1, some of which will overlap. Much care and patience is required to analyse the complete spectrum.

1.3. Intensity of Vibrational–Electronic Spectra: the Franck–Condon Principle. Although quantum mechanics imposes no restrictions on the change in the vibrational quantum number during an electronic transition, the vibrational lines in a progression are not all observed to be of the same intensity. In some spectra the (0, 0) transition is the strongest, in others the intensity increases to a maximum at some value of v', while in yet others only a few vibrational lines with high v' are seen, followed by a continuum. All these types of spectra are readily explicable in terms of the *Franck–Condon principle* which states that *an electronic transition takes place so rapidly that a vibrating molecule does not change its internuclear distance appreciably during the transition.*

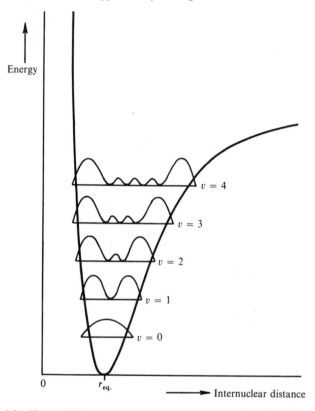

Fig. 6.2: The probability distribution for a diatomic molecule according to the quantum theory. The nuclei are most likely to be found at distances apart given by the maxima of the curve for each vibrational state.

We have already seen in Chapter 3 how the energy of a diatomic molecule varies with internuclear distance (cf. Fig. 3.3, p. 64). We recall that this figure, the Morse curve, represents the energy when one atom is considered fixed on the $r=0$ axis and the other is allowed to oscillate between the limits of the curve. Classical theory would suggest that the oscillating atom would spend most of its time *on* the curve at the turning point of its motion, since it is moving most slowly there; quantum theory, while agreeing with this view for high values of the vibrational quantum number, shows that for $v=0$ the atom is most likely to be found at the *centre* of its motion, i.e. at the equilibrium internuclear distance $r_{eq.}$.   For $v=1$, 2, 3,... the most probable positions steadily approach the extremities until, for high v, the quantal and classical pictures merge. This behaviour is shown in Fig. 6.2 where we plot the probability distribution in each vibrational state against internuclear distance. Those who have studied quantum mechanics will realize that Fig. 6.2 shows the variation of ψ^2 with internuclear distance, where ψ is the vibrational wave function.

If a diatomic molecule undergoes a transition into an upper electronic state in which the excited molecule is stable with respect to dissociation into its atoms, then we can represent the upper state by a Morse curve similar in outline to that of the ground electronic state.   There will probably (but not necessarily) be differences in such parameters as vibrational frequency, equilibrium internuclear distance, or dissociation energy between the two states, but this simply means that we should consider each excited molecule as a new, but rather similar, molecule with a different, but also rather similar, Morse curve.

Figure 6.3 shows three possibilities.   In (*a*) we show the upper electronic state having the same equilibrium internuclear distance as the lower.   Now the Franck–Condon principle suggests that a transition occurs *vertically* on this diagram, since the internuclear distance does not change, and so if we consider the molecule to be initially in the ground state both electronically (ε'') and vibrationally ($v''=0$), then the most probable transition is that indicated by the vertical line in Fig. 6.3(*a*).   Thus the strongest spectral line of the $v''=0$ progression will be the (0, 0).   However, the quantum theory only says that the *probability* of finding the oscillating atom is greatest at the equilibrium distance in the $v=0$ state—it allows some, although small, chance of the atom being near the extremities of its vibrational motion.   Hence there is some chance of the transition starting from the ends of the $v''=0$ state and finishing in the $v'=1$, 2, etc., states.   The (1, 0), (2, 0), etc., lines diminish rapidly in intensity, however, as shown at the foot of Fig. 6.3(*a*).

In Fig. 6.3(*b*) we show the case where the excited electronic state has a *slightly* greater internuclear separation than the ground state. Now a vertical transition from the $v'' = 0$ level will most likely occur into the upper vibrational state $v' = 2$, transitions to lower and higher v' states being less likely; in general the upper state most

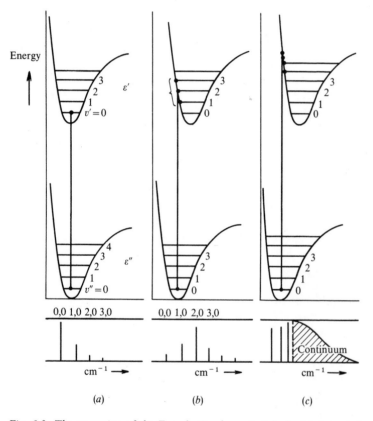

Fig. 6.3: *The operation of the Franck–Condon principle for* (a) *internuclear distances equal in upper and lower states,* (b) *upper state internuclear distance a little greater than that in the lower state, and* (c) *upper state distance considerably greater.*

probably reached will depend on the difference between the equilibrium separations in the lower and upper state.   In Fig. 6.3(*c*) the upper state separation is drawn as *considerably* greater than that in the lower state and we see that, firstly, the vibrational level to which a transition takes place has a high v' value. Further, transitions can now occur to a state where the excited molecule has energy in excess of its own dissociation energy.   From such states the molecule will dissociate without any vibrations and, since the atoms which are formed may take up any value of kinetic energy, the transitions are not quantized and a continuum results.   This is shown at the foot of the figure.   We consider the phenomenon of dissociation more fully in the next section.

The situation is rather more complex for emission spectra or for absorption from an excited vibrational state, for now transitions take place from *both ends* of the vibrational limits with equal probability; hence each progression will show two maxima which will coincide only if the equilibrium separations are the same in both states.

1.4. Dissociation Energy and Dissociation Products.   Fig. 6.4(a) and (b) show two of the ways in which electronic excitation can lead to dissociation (a third way called *predissociation*, will be considered in Section 1.7).   Part (a) of the figure represents the case, previously discussed, where the equilibrium nuclear separation in the upper

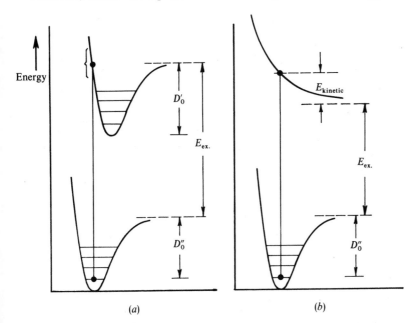

Fig. 6.4: Illustrating dissociation by excitation into (a) *a stable upper state, and* (b) *a continuous upper state.*

state is considerably greater than that in the lower.   The dotted line limits of the Morse curves represent the dissociation of the normal and excited molecule into atoms, the dissociation energies being D_0'' and D_0' from the $v = 0$ state in each case.   We see that the total energy of the dissociation products (i.e. atoms) from the upper state is greater by an amount called $E_{ex.}$ than that of the products of dissociation in the lower state.   This energy is the *excitation energy* of one (or rarely both) of the atoms produced on dissociation.

We saw in the previous section that the spectrum of this system consists of some vibrational transitions (quantized) followed by a continuum (non-quantized transitions) representing dissociation. The lower wavenumber limit of this continuum must represent just sufficient energy to cause dissociation and no more (i.e. the dissociation products separate with virtually zero kinetic energy) and thus we have

$$\bar{v}_{(\text{continuum limit})} = D_0'' + E_{ex.} \text{ cm}^{-1} \qquad (6.7)$$

and we see that we can measure D_0'', the dissociation energy, if we know $E_{ex.}$, the excitation energy of the products, whatever they may be.   Now, although the excitation energy of atoms to various electronic states is readily measurable by atomic spectroscopy (cf. Chapter 5), the precise *state* of dissociation products is not always obvious.   There are several ways in which the total energy $D_0'' + E_{ex.}$ may be separated into its components, however; here we shall mention just two.

Firstly, thermochemical studies often lead to an approximate value of D_0'' and hence, since $D_0'' + E_{ex.}$ is accurately measurable spectroscopically, a rough value for $E_{ex.}$ is obtained.   When the spectrum of the atomic products is studied, it usually happens that only one value of excitation energy corresponds at all well with $E_{ex.}$. Thus the state of the products is known, $E_{ex.}$ measured accurately and a precise value of D_0'' deduced.

Secondly, if more than one spectroscopic dissociation limit is found, corresponding to dissociation into two or more different states of products with different excitation energies, the separations between the excitation energies are often found to correspond closely with the separations between only one set of excited states of the atoms observed spectroscopically.   Thus the nature of the excited products and their energies are immediately known.

In Fig. 6.4(b) we illustrate the case in which the upper electronic state is *unstable*: there is no minimum in the energy curve and, as soon as a molecule is raised to this state by excitation, the molecule dissociates into products with total excitation energy $E_{ex.}$. The products fly apart with kinetic energy $E_{kinetic}$ which represents (as shown on the figure) the excess energy in the final state above that needed just to dissociate the molecule. Since $E_{kinetic}$ is not quantized the whole spectrum for this system will exhibit a continuum, the lower limit of which (if observable) will be precisely the energy $D_0'' + E_{ex.}$. As before, if $E_{ex.}$ can be found from a knowledge of the dissociation products, D_0'' can be measured with great accuracy.

We shall see in Section 2.1 what sort of circumstances lead to the minimum in the upper state (Fig. 6.4(*a*)) on the one hand, or the *continuous* upper state (Fig. 6.4(*b*)) on the other.

In many electronic spectra no continua appear at all—the internuclear distances in the upper and lower states are such that transitions near to the dissociation limit are of negligible probability —but it is still possible to derive a value for the dissociation energy by noting how the vibrational lines converge. We have already seen in Chapter 3 (cf. equation (3.12)), that the vibrational energy levels may be written:

$$\varepsilon_v = (v+\tfrac{1}{2})\bar{\omega}_e - x_e(v+\tfrac{1}{2})^2\bar{\omega}_e \text{ cm}^{-1} \qquad (6.8)$$

and so the separation between neighbouring levels, $\Delta\varepsilon$, is plainly:

$$\Delta\varepsilon = \varepsilon_{v+1} - \varepsilon_v$$
$$= \bar{\omega}_e\{1 - 2x_e(v+1)\} \text{ cm}^{-1}. \qquad (6.9)$$

This separation obviously decreases linearly with increasing v and the dissociation limit is reached when $\Delta\varepsilon \to 0$. Thus the maximum value of v is given by $v_{max.}$, where:

$$\bar{\omega}_e\{1 - 2x_e(v_{max.} + 1)\} = 0,$$

i.e.

$$v_{max.} = \frac{1}{2x_e} - 1. \qquad (6.10)$$

We recall that the anharmonicity constant, x_e, is of the order 10^{-2}, hence $v_{max.}$ is about 50.

We saw in Chapter 3, Section 1.3, that two vibrational transitions (in the infra-red) were sufficient to determine x_e and $\bar{\omega}_e$. Thus, an example given there for HCl yielded $\bar{\omega}_e = 2990$ cm^{-1}, $x_e = 0{\cdot}0174$. From equation (6.10) we calculate $v_{\text{max.}} = 27{\cdot}74$ and the next lowest integer is $v = 27$.   Replacing $v = 27$, $\bar{\omega}_e = 2990$ cm^{-1} and $x_e = 0{\cdot}0174$ into equation (6.8) gives the maximum value of the vibrational energy as 42,890 cm^{-1} or 122·6 kcal/mole.   This is to be compared with a more accurate value of 102·1 kcal/mole evaluated thermo-chemically.

The discrepancy between these two figures arises from two causes. Firstly, the infra-red data only allows us to consider two or three vibrational transitions (the fundamental plus the first and second overtones).   The electronic spectrum, as we have seen, shows many more vibrational lines (in fact the number is limited not by quantum restrictions, but by the Franck–Condon principle) and we shall get a better value of D_0'' if we make use of this extra data.   Secondly, we have assumed that equation (6.8) applies exactly even at high values of v; this is not true because cubic and even quartic terms become important at this stage.   Because of these, $\Delta\varepsilon$ decreases more rapidly than equation (6.9) suggests.

Both these points may be met if we plot the *separation* between vibrational transitions, $\Delta\varepsilon$, as observed in the electronic spectrum, against the vibrational quantum number.   Initially, equation (6.9) will apply quite accurately and the graph will be a straight line which may be extrapolated either to find $v_{\text{max.}}$ or, since the dissociation energy itself is simply the sum of all the increments $\Delta\varepsilon$ from $v = 0$ to $v = v_{\text{max.}}$, the *area* under the $\Delta\varepsilon$ vs. v graph gives this energy directly.   Such a linear extrapolation was first suggested by Birge and Sponer and is usually given their name.

On the other hand, if extensive data is available about a set of electronic–vibration transitions, the graph of $\Delta\varepsilon$ vs. v will, at high v, begin to fall off more sharply as cubic and quartic terms become significant.   In this case the most accurate determination of dissociation energy is obtained by extrapolating the smooth curve and finding the area beneath it.   Figure 6.5 shows this process for data on iodine vapour given by R. D. Verma, *J. Chem. Phys.*, **32**, 738 (1960).

In absorption spectra it is normally the series of lines originating at $v'' = 0$ which is observed (cf. Fig. 6.1).   Thus the convergence of the levels in the upper state and hence the dissociation energy of that state is normally found.   While this in itself is of great interest, particularly since molecules in excited states usually revert to the ground state within fractions of a microsecond, the dissociation energy in the ground state can be found quite easily provided, as before, the dissociation products and their excitation energy are known. Thus, in Fig. 6.4(a), if we know $E_{ex.}$ (from atomic spectroscopy), D_0' (from Birge–Sponer extrapolation) and, if we can measure

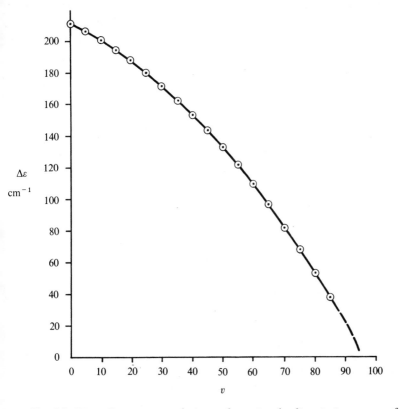

Fig. 6.5: *Birge–Sponer extrapolation to determine the dissociation energy of the iodine molecule,* I_2. *(Taken from the data of R. D. Verma, J. Chem. Phys.,* **32,** *738 (1960), by kind permission of the author.)*

the energy of the (0, 0) transition, either directly or by calculation from the observed energy levels, we have:

$$D_0'' = \text{energy of } (0,0) + D_0' - E_{ex.} \text{ cm}^{-1}. \qquad (6.11)$$

1.5. Rotational Fine Structure of Electronic–Vibration Transitions. So far we have seen that the electronic spectrum of a diatomic

14

molecule consists of one or more series of convergent lines constituting the vibrational coarse structure on each electronic transition.   Normally each of these "lines" is observed to be broad and diffuse or, if the resolution is sufficiently good, each appears as a cluster of many very close lines.   This is, of course, the rotational fine structure.

To a very good approximation we can ignore centrifugal distortion and we have the energy levels of a rotating diatomic molecule (cf. equations (2.11 and 2.12)) as:

$$\varepsilon_{\text{rot.}} = \frac{h}{8\pi^2 Ic} J(J+1) = BJ(J+1) \text{ cm}^{-1} \quad (J = 0, 1, 2, \ldots) \quad (6.12)$$

where I is the moment of inertia, B the rotational constant and J the rotational quantum number.   Thus, by the Born–Oppenheimer approximation, the total energy (excluding kinetic energy of translation) of a diatomic molecule is:

$$\varepsilon_{\text{total}} = \varepsilon_{\text{elec.}} + \varepsilon_{\text{vib.}} + BJ(J+1) \text{ cm}^{-1}. \quad (6.13)$$

Changes in the total energy may be written:

$$\Delta\varepsilon_{\text{total}} = \Delta\{\varepsilon_{\text{elect.}} + \varepsilon_{\text{vib.}}\} + \Delta\{BJ(J+1)\} \text{ cm}^{-1} \quad (6.14)$$

and the wavenumber of a spectroscopic line corresponding to such a change becomes simply:

$$\bar{v}_{\text{spect.}} = \bar{v}_{(v', v'')} + \Delta\{BJ(J+1)\} \text{ cm}^{-1} \quad (6.15)$$

where we write $\bar{v}_{(v', v'')}$ to represent the wavenumber of an electronic-vibrational transition.   This plainly corresponds to any *one* of the transitions, e.g. (0, 0) or (1, 0), etc., considered in previous sections. Here we are mainly concerned with $\Delta\{BJ(J+1)\}$.

The selection rule for J depends upon the type of electronic transition undergone by the molecule.   We shall discuss these in more detail in Section 2.2; for the moment we must simply state that if both the upper and lower electronic states are $^1\Sigma$ states (i.e. states in which there is no electronic angular momentum about the internuclear axis), this selection rule is:

$$\Delta J = \pm 1 \text{ only, for } {}^1\Sigma \to {}^1\Sigma \text{ transitions,} \quad (6.16)$$

whereas for all other transitions (i.e. provided either the upper or the lower states (or both) have angular momentum about the bond axis) the selection rule becomes:

$$\Delta J = 0, \text{ or } \pm 1. \tag{6.17}$$

For both these cases there is the added restriction that a state with $J=0$ cannot undergo a transition to another $J=0$ state:

$$J = 0 \nleftrightarrow J = 0. \tag{6.18}$$

Thus we see that for transitions between $^1\Sigma$ states, P and R branches only will occur, while for other transitions Q branches will appear in addition.

We can expand equation (6.15) as follows:

$$\bar{\nu}_{\text{spect.}} = \bar{\nu}_{(v',v'')} + B'J'(J'+1) - B''J''(J''+1) \text{ cm}^{-1} \tag{6.19}$$

where B' and J' refer to the upper electronic state, B'' and J'' to the lower. When we considered vibration–rotational spectra in Chapter 3, we saw (cf. Section 4 of that chapter) that the difference between B-values in different vibrational levels was very small and could be ignored except in explaining finer details of the spectra. But this is by no means the case in electronic spectroscopy: here we have seen, when discussing the Franck–Condon principle in Section 1.3, that equilibrium internuclear distances in the lower and upper electronic states may differ considerably, in which case the moments of inertia, and hence B-values, in the two states will also differ. We cannot say *a priori* which of the two B-values will be greater. Quite often the electron excited is one of those forming the bond between the nuclei; if this is so, the bond in the upper state will be weaker and probably longer (cf. Fig. 6.3(b) or (c)) so that the equilibrium moment of inertia increases during the transition and B decreases. Thus $B' < B''$. The reverse is sometimes true, however, e.g. when the electron is excited from an antibonding orbital (see Section 2.2).

We can discuss the rotational fine structure quite generally by applying the selection rules of equations (6.16, 6.17 and 6.18) to the expression for spectral lines, equation (6.19). We may note, in passing, that the treatment given here for the P and R branch lines is identical with that given in Section 4 of Chapter 3 for the vibration–rotation spectrum, except that there we were concerned with B_0 and B_1—B-values in lower and upper *vibrational* states. Here our concern is with B-values in lower and upper *electronic* states, B'' and B', and we also consider the formation of a Q branch.

Taking the P, R and Q branches in turn:
(i) P branch: $\Delta J = -1$, $J'' = J' + 1$

$$\Delta\varepsilon = \bar{v}_P = \bar{v}_{(v', v'')} - (B' + B'')(J' + 1) + (B' - B'')(J' + 1)^2 \text{ cm}^{-1},$$

$$\text{where } J' = 0, 1, 2, \ldots \quad (6.20a)$$

(ii) R branch: $\Delta J = +1$, $J' = J'' + 1$

$$\Delta\varepsilon = \bar{v}_R = \bar{v}_{(v', v'')} + (B' + B'')(J'' + 1) + (B' - B'')(J'' + 1)^2 \text{ cm}^{-1},$$

$$\text{where } J'' = 0, 1, 2, \ldots \quad (6.20b)$$

These two equations can be combined into:

$$\bar{v}_{P, R} = \bar{v}_{(v', v'')} + (B' + B'')m + (B' - B'')m^2 \text{ cm}^{-1},$$

$$\text{where } m = \pm 1, \pm 2, \ldots \quad (6.20c)$$

positive m values comprising the R branch (i.e. corresponding to $\Delta J = +1$) and negative values the P branch ($\Delta J = -1$). Note that m cannot be zero (this would correspond in, e.g. the P branch, to $J' = -1$ which is impossible) so that no lines from the P and R branch appears at the *band origin* $\bar{v}_{(v', v'')}$. We draw the appearance of the R and P branches separately in Fig. 6.6(a) and (b) respectively, taking a 10% difference between the upper and lower B-values and choosing $B' < B''$. Note that, *with this choice*, P branch lines occur on the *low* wavenumber side of the band origin and the spacing between the lines increases with m. On the other hand the R branch appears on the *high* wavenumber of the origin and the line spacing decreases rapidly with m—so rapidly that the lines eventually reach a maximum wavenumber and then begin to return to low wavenumbers with increasing spacing.* It will be remembered that in Chapter 3, Section 4, a similar decrease in spacing was observed in the R branch but this was much too slow for a convergence limit to be reached; the rapid convergence here is due simply to the magnitude of $B' - B''$. The point at which the R branch separation decreases to zero is termed the *band head*.

(iii) Q branch: $\Delta J = 0$, $J' = J''$

$$\Delta\varepsilon = \bar{v}_Q = \bar{v}_{(v', v'')} + (B' - B'')J'' + (B' - B'')J''^2 \text{ cm}^{-1},$$

$$\text{where } J'' = 1, 2, 3, \ldots. \quad (6.21)$$

---

* The returning lines of the R branch coincide with earlier lines if equation (6.20b) is obeyed exactly. For real molecules cubic and quartic terms become important at high values of m.

Note that here $J'' = J' \neq 0$ since we have the restriction shown in equation (6.18). Thus again *no line will appear at the band origin*. We sketch the Q branch in Fig. 6.6(c), again for $B' < B''$ and a 10% difference between the two. We see that the lines lie to *low* wavenumber of the origin and their spacing increases. The first few lines of this branch are not usually resolved.

The complete rotational spectrum is shown in Fig. 6.6(d). We have seen in Chapter 2, Section 3.2, that many rotational levels are

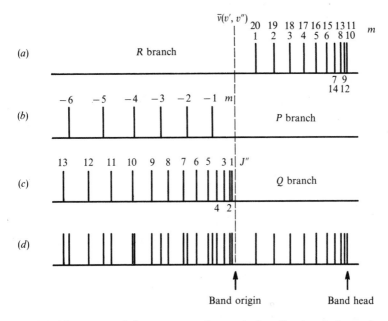

Fig. 6.6: *The rotational fine structure of a particular vibrational–electronic transition for a diatomic molecule. The R, P and Q branches are shown separately at (a), (b) and (c) respectively, with the complete spectrum at (d).*

populated even at room temperature; consequently, a large number of the P and R (and Q, where appropriate) lines will appear in the spectrum with comparable intensity. The spectrum is usually dominated by the band head, since here several of the R branch lines crowd together; for this reason, the Q branch is not very apparent if it occurs.

In the situation we have been discussing ($B' < B''$), the band head appears in the R branch on the high wavenumber side of the origin;

such a band is said to be *degraded* (or *shaded*) *towards the red*—i.e. the tail of the band where the intensity falls off points towards the red (low frequency) end of the spectrum. If, on the other hand, $B' > B''$, then all our previous arguments are reversed. Briefly: (*i*) the Q branch spreads to *high* wavenumber, (*ii*) the R branch (still, of course, on the *high* wavenumber side) consists of a series of lines with *increasing* separation and (*iii*) the band head appears in the P branch to *low* frequency of the origin. Such a band is *shaded to the violet*.

Normally, all the vibrational bands in any one electronic transition (e.g. the set of bands shown as a line spectrum in Fig. 6.1) are shaded in the same direction, while different electronic transitions in the same molecule may well show different shadings. Thus, observation of the shading may assist in the analysis of a complex spectrum. However, it may happen that different shadings are observed in bands belonging to the same electronic transition. This is because the B' and B'' values are not altogether independent of the vibrational state (as we have already seen in Chapter 3, Section 4) so that, if $B' - B''$ is small, it may reverse sign for some higher vibrational levels. This behaviour is observed, for example, in the molecular fragment AlF, but is rare.

1.6. The Fortrat Diagram. We may rewrite the expressions for the P, R and Q lines, equations (6.20c and 6.21), with *continuously variable* parameters p and q:

$$\bar{\nu}_{P,R} = \bar{\nu}_{(v',v'')} + (B' + B'')p + (B' - B'')p^2 \qquad (6.22a)$$

$$\bar{\nu}_Q = \bar{\nu}_{(v',v'')} + (B' - B'')q + (B' - B'')q^2 \qquad (6.22b)$$

when we see that they each represent a parabola, p taking both positive and negative values, while q is positive only. We sketch these parabolae in Fig. 6.7 choosing, as before, $B' < B''$ and a difference of 10% between them, and labelling regions of positive p with $\bar{\nu}_R$ and negative p with $\bar{\nu}_P$. These parabolae are usually referred to as the Fortrat parabolae. If we now illustrate the fact that p and q may in fact take only integral values (but not zero) by drawing circles round the allowed points on the parabolae, we can then read off the $\bar{\nu}$-values of the spectral lines directly from the graph. We show at the foot of the figure the first few lines of each branch with dotted leader lines connecting spectrum and Fortrat diagram at intervals.

A useful property of the Fortrat diagram is that the band-head is plainly at the vertex of the P, R parabola. We may calculate the position of the vertex by differentiation of equation (6.22a):

$$\frac{d\bar{v}_{P,R}}{dp} = B' + B'' + 2(B' - B'')p = 0$$

or

$$p = -\frac{B' + B''}{2(B' - B'')} \quad \text{for band head.} \tag{6.23}$$

Thus if $B' < B''$ (upper state has longer equilibrium bond-length) the band head occurs at *positive p* values (i.e. in the R branch), the line at

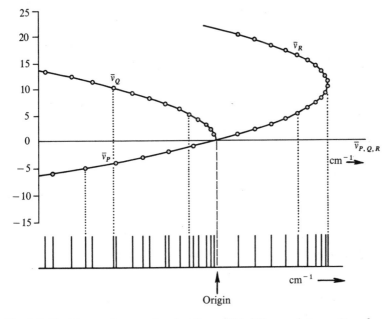

Fig. 6.7: *The Fortrat diagram sketched for a 10% difference between B' and B'' (with B' < B''); the spectrum illustrated at the foot is identical with that of Fig. 6.6(d).*

maximum wavenumber being given by the nearest positive integer to p. Conversely, for $B' > B''$ the band head occurs in the region of p negative, i.e. in the P branch. A simple calculation shows that for a 10% difference between B' and B'' the band head occurs at $p \approx 10$.

1.7. Predissociation.   If a large number of vibrational transitions are observed for a particular molecule, it sometimes happens that the vibrational and rotational structure are quite distinct within a progression for large and small changes in the vibrational quantum number, but either the rotational structure is blurred or a complete continuum is observed for intermediate changes. A diagram showing the appearance of such a band is sketched in Fig. 6.8.   A

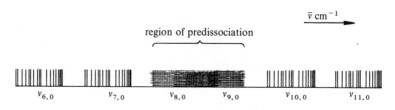

Fig. 6.8: Diagrammatic illustration of the appearance of predissociation. The rotational fine structure is clearly defined for vibrational transitions both above and below the predissociation region, but in this region the fine structure becomes blurred and lost.

continuum at high wavenumber would correspond to ordinary dissociation (cf. Section 1.4) but the central continuum, occurring at energies well below the true dissociation limit, is referred to as *predissociation.*

Predissociation can arise when the Morse curves of a particular molecule in two different excited states intersect; one such possibility is shown in Fig. 6.9.   One of the excited states is stable, since it has a minimum in the curve, the other is continuous.   Some of the vibrational levels are also shown, and let us suppose a transition takes place from some lower state into the vibrational levels shown bracketed on the left.   Now if a transition takes place into the levels labelled *a, b* or *c,* a normal vibrational–electronic spectrum occurs complete with rotational fine structure; two such bands appear at the left of our previous Fig. 6.8.   If the transition is to levels *d* or *e* there is a possibility that the molecule will "cross over" on to the continuous curve and thus dissociate.   In general, transition from one curve to another in this way (a so-called *radiationless transfer* since no energy is absorbed or emitted in the process) is faster than the time taken by the molecule to rotate ($\sim 10^{-10}$ sec) but usually slower than the vibrational time ($\sim 10^{-13}$ sec).   Thus predissociation will occur before the molecule rotates (and thus all rotational

fine structure will be destroyed in the spectrum), while the vibrational structure is usually not destroyed.   If the cross-over is faster
than the vibrational time, then a complete continuum will occur in
the spectrum as shown in Fig. 6.8.

On the other hand, transitions into levels f, g, h,... will give rise
to a normal vibrational–electronic spectrum including rotational
fine structure once more.   As we have seen previously (Section 1.3)

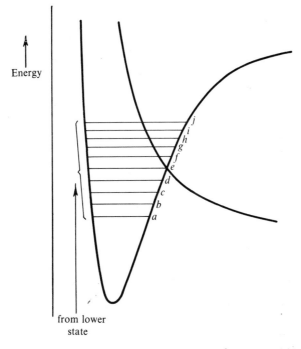

Fig. 6.9: *Showing the occurrence of predissociation during transitions into a
stable upper state intersected by a continuous state.*

the molecule spends most time at the extreme ends of its vibrational
motion when v is large, and very little time in between.   When
moving in the vibrational states f, g,..., the molecule spends insufficient time near the cross-over point for appreciable dissociation
to occur and a normal spectrum results.

1.8. Diatomic Molecules: a Summary.   When the rotational fine
structure of electronic spectra can be resolved—as it normally can

for diatomic molecules—we see that a great deal of useful information becomes available. We can immediately determine the rotational constant, and hence calculate the moment of inertia and bond length, for both the lower and the upper electronic states. Isotopic species in the molecule will cause a slight difference in the rotational constant, so such isotopes may be detected, and their concentrations measured from the band intensity. Equally, the vibrational levels of the electronic states can be determined from the position of band origins; these lead to the evaluation of fundamental vibration frequencies, of bond force constants and, perhaps, of dissociation energies too. The latter, however, are more accurately determined if a continuum is observed at the end of a band spectrum.

Where data obtained from such spectra can be checked independently, e.g. by microwave or infra-red spectroscopy, by X-ray or neutron diffraction, or by thermochemical methods, perfectly satisfactory agreement is found. Thus we can use electronic spectroscopic methods with great confidence to determine bond lengths and strengths in those molecules to which such independent methods are not applicable, e.g. homonuclear molecules which have no pure rotation or vibration spectra.

Probably the most important application of this type of spectroscopy is to the study of excited states and unstable radicals. Thus we have seen that B-values and dissociation energies are obtained for both the upper and lower electronic states—and data for the upper state are not obtainable by other means. Further, considerable amounts of energy are involved in the production of electronic spectra, and complex molecules are frequently disrupted into fragments during the process, the fragments, or free-radicals, normally being very short-lived. Examples are legion, a few of the more important diatomic ones being CH, NH, C_2, OH, CN, etc. Spectra arising from these radicals can be recognized and studied, leading to the determination of bond lengths, force constants, dissociation energies, etc. Further, if the variation of the intensity of such spectra over short periods of time is studied—as in the techniques of flash photolysis—information can be obtained about the rate at which the radicals are produced and destroyed. Since the length of time during which some radicals have an independent existence is measured in microseconds or less, it is remarkable that many such "diatomic molecules" are as well characterized as, for example, the rather more stable H_2 or CO.

2. Electronic Structure of Diatomic Molecules

2.1. Molecular Orbital Theory. Several theories have been suggested to account for the formation of molecules from atoms. All, if taken to a sufficiently high degree of approximation, seem to agree with observed data, but the calculation involved is so extensive that complete agreement is seldom reached and then only in the simplest examples. Here we shall discuss just one of these theories—the *molecular orbital theory*; we choose this, not because it is better or simpler than others (such considerations depend upon the particular problem in hand and are, in any case, largely subjective), but because it gives a convenient pictorial representation of molecule formation which is particularly suited to the discussion of electronic transitions, and because the ideas it uses are entirely analogous to those of atomic structure which we have discussed in the previous chapter.

Thus we have seen that electrons in atoms do not occupy space haphazardly or have arbitrary energies, but that their distribution and energy are governed by well-defined natural laws. These characteristics may be calculated from the Schrödinger equation and expressed in terms of a three-dimensional wave function, or orbital, ψ, which depends on the values of three quantum numbers, n, l and m (or l_z); the spin of the electron also contributes to the energy. Definite rules determine which orbitals are occupied in the ground state and what transitions may take place between orbitals.

Molecular orbital theory supposes orbitals to extend about, and embrace, *two or more nuclei*, the shape and energy of these orbitals being calculable from the Schrödinger equation in terms of three quantum numbers. Essentially the same rules (i.e. lowest energy first, maximum of two (paired) electrons per orbital, parallel spins in degenerate orbitals) apply to their filling as to the filling of atomic orbitals.

The situation is relatively simple for diatomic molecules where the molecular orbital embraces two nuclei only and we shall discuss these molecules in some detail first. The extension to polyatomic molecules will be outlined in Section 3.

2.2. The Shapes of Some Molecular Orbitals. As in atomic orbital theory (cf. Chapter 5, Section 1.2) the shape of a molecular orbital is the space within which an electron belonging to that orbital will spend 95% (or some other arbitrary fraction) of its time.

While detailed computation of these shapes from the Schrödinger theory may be extremely difficult, a very good qualitative idea of their approximate shape may be obtained by considering molecular orbitals to be made up of sums and differences of the atomic orbitals of the constituent atoms—the so-called *linear combination of atomic orbitals* (LCAO) approximation. Thus for a diatomic molecule we could imagine the formation of two different molecular orbitals whose wave functions would be:

$$\psi_{mo.} = \psi_1 + \psi_2 \quad \text{or} \quad \psi_{mo.} = \psi_1 - \psi_2 \tag{6.24}$$

where ψ_1 and ψ_2 are the relevant atomic orbitals of the two atoms. Note that the function $\psi_2 - \psi_1$ is identical with $\psi_1 - \psi_2$, since it is $\psi_{mo.}^2$. which represents the probability of finding an electron in a particular place.

Let us consider the hydrogen molecule, H_2, as an example; the obvious atomic orbitals to use are the $1s$ orbitals of each atom. Figure 6.10(a) shows the situation:

$$\psi_{H_2} = \psi_{1s} + \psi_{1s}. \tag{6.25}$$

We recall from the previous chapter (equation (5.2)) that ψ_{1s} is everywhere positive in value and so, where the atomic orbitals overlap the value of ψ_{H_2} will be increased. This suggests (and detailed calculation confirms) that the molecular orbital of equation (6.25) is a simple ellipsoid, symmetrical in shape. The increase of electronic charge between the nuclei acts as a sort of cement keeping the nuclei together and thus this orbital represents the formation of a bond between the atoms. It is called a *bonding orbital* and given the label 1σ since it is produced from two $1s$ orbitals (we shall see later that σ has a similar significance regarding the orbital angular momenta of molecular electrons as s has for atoms).

On the other hand, Fig. 6.10(b) shows the situation for

$$\psi_{H_2} = \psi_{1s} - \psi_{1s}. \tag{6.26}$$

Since, again, ψ_{1s} is everywhere positive, where the two separate ψ_{1s} orbitals overlap, they will cancel each other. Thus between the nuclei $\psi_{mo.}$ will be zero, while it will be positive near one nucleus and negative near the other (remember that it is ψ^2 which determines probability and this is, in either case, positive). Now, however, the shape of the molecular orbital shows that electronic charge, far from being concentrated *between* the nuclei, is greatest *outside* them; thus the nuclear repulsion is enhanced and the orbital is described as *antibonding*. It leads to a state of higher energy than two separate atoms and is labelled $1\sigma^*$, the asterisk representing high energy. Figure 6.10 also shows, at the extreme right, an end-view of these orbitals. They are both seen to have cylindrical symmetry about the bond axis; it is this property which leads to their both being described as σ orbitals although in other respects they have quite different appearances.

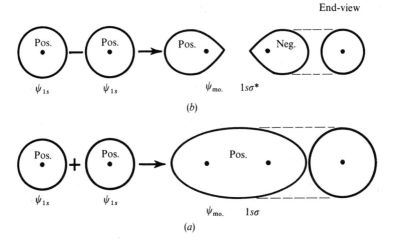

Fig. 6.10: *The formation of* (a) *a bonding* $1s\sigma$ *orbital and* (b) *an antibonding* $1s\sigma^*$ *orbital from two atomic 1s orbitals.*

Another facet of orbital symmetry should be mentioned here. If the molecule considered is homonuclear (i.e. made of two *identical* atoms) then the mid-point of the bond between them is a *centre of symmetry*—starting from any point in the molecule, on or off the internuclear axis, exactly similar surroundings are encountered by proceeding to the point diagonally opposite the centre. Such a process is known as *inversion*, and such molecular properties as electron density, force fields, etc., are quite unchanged by inversion. However, we note that ψ (as opposed to ψ^2) may or may not be changed *in sign* by inversion. Thus inversion of the 1σ molecular orbital of Fig. 6.10(a) plainly causes no change in ψ since it is everywhere positive; this orbital, in which ψ is completely symmetrical, is described as *even* and usually given the symbol g (German:

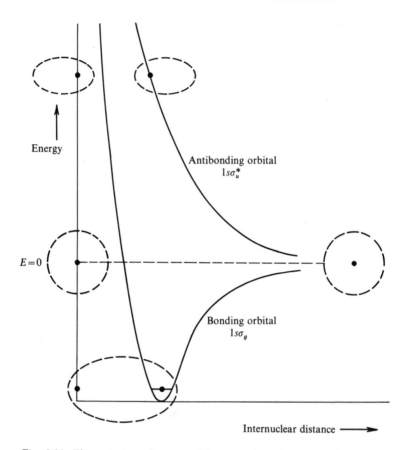

Fig. 6.11: The variation of energy with internuclear distance in the bonding and antibonding orbitals, $1s\sigma_g$ and $1s\sigma_u^*$.

gerade = even) as a suffix: $1\sigma_g$. On the other hand the $1\sigma^*$ orbital in (*b*) of the figure is antisymmetrical, since inversion reverses the sign of ψ. This orbital is thus *odd* and given the subscript u, $1\sigma_u$ or $1\sigma_u^*$ (German: *ungerade* = odd). In the case of molecular hydrogen, then, the bonding orbital is even, the antibonding is odd; this situation may be reversed for other molecular orbitals as we shall see.

If the molecule is heteronuclear (e.g. CO, HCl, etc.) then there is no centre of symmetry and the odd–even classification of orbitals does not arise.

Before turning to the shapes of other molecular orbitals, it is instructive to consider how the energy of the $1\sigma_g$ and $1\sigma_u^*$ orbitals varies with the distance between the nuclei. This variation may be calculated from the Schrödinger equation and the result is shown in Fig. 6.11. The $1\sigma_g$, the bonding orbital, shows a typical Morse curve for a diatomic molecule, the minimum in the curve showing that a bond is formed between the atoms. The $1\sigma_u^*$, on the other hand, shows no minimum, but is one of the "continuous" curves already discussed in Section 1.4 of this chapter. In this case the dissociation limits of the two curves coincide since the dissociation products are identical—two hydrogen atoms. The relationship of the orbitals sketched in Fig. 6.10 to the energy curve is shown by superimposing the molecular orbitals at their appropriate inter-nuclear distance and the separate atomic orbitals at a large distance.

Two $2s$ atomic orbitals can form $2\sigma_g$ and $2\sigma_u^*$ bonding and anti-bonding orbitals with identical shape to (but larger and with higher energy than) the $1\sigma_g$ and $1\sigma_u^*$ orbitals. Two $2p$-type orbitals can overlap in two different ways depending on their relative orientation. If we label the internuclear axis the z-direction, then we may consider first the $2p$ orbital which lies along this axis for each atom, i.e. the $2p_z$ orbitals. Now the expression for the wave function of a $2p_z$ orbital has the form $\psi_{2p_z} = zf(r)$ where $f(r)$ is a positive function of distance from the nucleus. We see then, that for $+z$ directions ψ is also positive, while it is negative for $-z$. The two lobes of a $2p$ orbital thus have opposite signs of ψ (although, of course, ψ^2 is everywhere positive).

We draw these orbitals and indicate their signs on the left of Fig. 6.12 and it is evident that in the molecular orbital

$$\psi_{\text{mo.}} = \psi_{2p_z} + \psi_{2p_z} \tag{6.27}$$

the electron density between the nuclei is cancelled to zero and the orbital will have the shape shown at the top right of the figure. Similarly in

$$\psi_{\text{mo.}} = \psi_{2p_z} - \psi_{2p_z} \tag{6.27a}$$

the electron density increases between the nuclei and the shape is shown at the bottom right of the figure. Plainly the latter is *bonding*, and consideration of its symmetry shows that it is *even* (*g*), while the former is plainly *antibonding* and *odd* (*u*) in character. The end-view of both is the same, however, and shows symmetry about the bond axis; for this reason both are referred to as σ

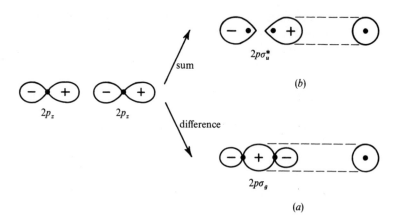

Fig. 6.12: *The formation of* (a) *a bonding* $2p\sigma_g$ *and* (b) *an antibonding* $2p\sigma_u^*$ *orbital from two atomic* $2p_z$ *orbitals, where the z-axis is taken as the internuclear axis.*

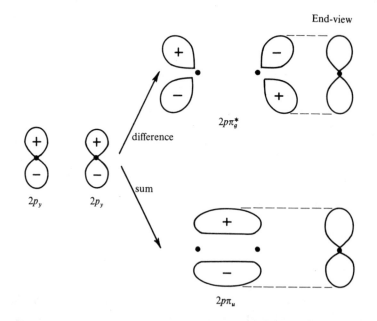

Fig. 6.13: *The formation of bonding* $(2p\pi_u)$ *and antibonding* $(2p\pi_g^*)$ *orbitals from two atomic* $2p_y$ *orbitals, the z-axis being the internuclear axis. Atomic* $2p_x$ *orbitals would form identical molecular orbitals except that all lobes would be rotated through a right angle about the z-axis.*

orbitals and they may be labelled $2p\sigma_g$ and $2p\sigma_u^*$ respectively to indicate their origin.

The overlap of two $2p_y$ orbitals is shown in Fig. 6.13 ($2p_x$ are exactly similar but rotated out of the plane of the paper through 90°). The summed orbitals, which are *bonding*, are sketched at bottom right and we see that the molecular orbital formed consists of two streamers, one above and one below the nuclei. The end-view of this orbital is shown at the extreme right; it has the appearance of an atomic p orbital and hence it is labelled π. In this case the bonding orbital is evidently *odd*, so we have a $2p\pi_u$ state.

On the other hand, if the atomic orbitals are subtracted the orbital picture shown at top right of Fig. 6.13 is produced. This has a similar end-view to $2p\pi_u$, is *antibonding* and *even*, hence it is labelled π_g^*.

More complex orbitals exist—δ, ϕ, etc., formed by interaction between d, f, etc., atomic orbitals—but they need not concern us; the simple molecules with which we shall deal use σ and π orbitals only. We do need to know, however, the order of increasing energy of the molecular orbitals so far discussed so that we can consider the ground states of some atoms. Briefly:

$$1\sigma_g < 1\sigma_u^* < 2\sigma_g < 2\sigma_u^* < 2p\sigma_g < \pi_u < \pi_g^* < 2p\sigma_u^*. \quad (6.28)$$

(For some light molecules π_u comes below $2\sigma_u$ and $2p\sigma_g$ in energy, but this is not important to the argument.) Using this list of orbital energies and the Pauli principle (not more than two electrons to each orbital) we can build up the electronic configurations of simple molecules. Some examples follow.

Firstly, hydrogen, H_2: in the ground state of this molecule the two $1s$ atomic electrons, one from each atom, can both occupy the molecular $1\sigma_g$ provided their spins are opposed. The energy of electrons in this state, as we have seen from Fig. 6.11, is lower than that of two separate atoms, hence the molecule is stable. We can write its configuration: $1\sigma_g^2$.

Next, helium, He_2: if this molecule were to form it would have a total of four electrons to place into orbitals, two from each atom. Two of these only could go into $1\sigma_g$, the other two would have to go into $1\sigma_u^*$. But we can see from Fig. 6.11 that more energy would be *absorbed* by the latter than evolved by the former, hence the molecule is unstable with respect to the atoms.

15

216   ELECTRONIC SPECTROSCOPY OF MOLECULES

Finally, nitrogen, N_2: this is a more complicated example. The two atoms each have an electronic configuration $1s^2 2s^2 2p_x^1 2p_y^1 2p_z^1$ (cf. Chapter 5, Table 5.2), so the molecular orbitals formed during interaction, and the number of electrons each would contain is: $1\sigma_g^2 1\sigma_u^2 2\sigma_g^2 2\sigma_u^2 2p\sigma_g^2 \pi_{u(x)}^2 \pi_{u(y)}^2$ (where we omit the asterisk to avoid confusion). As an approximation we can allow the bonding and antibonding contributions of the pair $1\sigma_g^2$ and $1\sigma_u^2$ to cancel, and similarly with $2\sigma_g^2$ and $2\sigma_u^2$. We are left, then, with three pairs of electrons in bonding orbitals, the $2p\sigma_g$, $\pi_{u(x)}$ and $\pi_{u(y)}$, so we conclude that the molecule is triply-bonded.

2.3. Electronic Angular Momentum in Diatomic Molecules; Classification of States.   We found in Chapter 5, Section 2, that the total energy of an electron, while depending mainly on its average distance from the nucleus (represented by the quantum number n) also depends on its orbital and spin angular momenta (quantum numbers l and s) and on the way in which these are coupled together (quantum number j). For several electrons in an atom we found that their separate energies can be combined in different ways to produce a variety of states; simple rules allow the ground state to be predicted in any particular case.

Much the same comments apply to electrons in molecules. Thus a single electron in a molecule has a quantum number n specifying the size of its orbital and mainly determining its energy, and a number l specifying its orbital angular momentum. Small letters $s, p, d, \ldots$ are used, as before, to designate l-values of 0, 1, 2, ... However, it will be remembered that in order to discuss the components of l we required to invoke some reference direction called the z-direction; in a diatomic molecule a reference direction is already quite obviously specified—the internuclear axis, or bond—and it would be perverse (not to say *wrong*) to discuss the components of l along any other direction. Furthermore, a force-field exists along this direction due to the presence of two nuclear charges; therefore different l-components are not degenerate but represent *different energies*.

The axial component of orbital angular momentum is of more importance in molecules than the momentum itself and for this reason it is given the special symbol λ. Formally $\lambda \equiv |l_z|$, so that λ takes *positive* integral values or is zero, and we designate the λ-state of an electron in a molecule by using the small Greek letters corresponding to the $s, p, d, \ldots$ of atomic nomenclature. Thus we have:

$$\text{for} \quad l_z = 0, \pm 1, \pm 2, \pm 3, \ldots$$

$$\lambda = 0, 1, 2, 3, \ldots$$

and symbol is: $\sigma, \pi, \delta, \phi, \ldots$

Since λ has positive values only, each λ-state with $\lambda > 0$ is *doubly degenerate*, because it corresponds to l_z being both positive and negative. The significance of λ is that the *axial component of orbital angular momentum* $= \lambda h/2\pi$ or λ a.m. units.

The total orbital angular momentum of several electrons in a molecule can be discussed, as for atoms, in terms of the quantum number $L = \Sigma l, \Sigma l - 1$, etc., with $\mathbf{L} = \sqrt{L(L+1)}h/2\pi$, but again the axial component, denoted by Λ, is of greatest significance. Since, by definition, all individual λ_i lie along the internuclear axis, their summation is particularly simple. We have

$$\Lambda = |\Sigma\lambda_i| \tag{6.29}$$

and states are designated by capital Greek letters Σ, Π, Δ, etc., for $\Lambda = 0, 1, 2, \ldots$. We must take into account, when using equation (6.29) that the individual λ_i may have the same or opposite directions and all possible combinations which give a positive Λ should be considered. Thus for a π and a δ electron ($\lambda_1 = 1$, $\lambda_2 = 2$) we could have $\Lambda = 1$ or 3 (but not -1), i.e. a Π or a Φ state.

Electron *spin* momentum, on the other hand, is not greatly affected by the electric field of the two nuclei—we say the spin–axial coupling is weak, whereas the orbital–axial coupling is usually strong. Normally, therefore, we use the same notation for electronic spin in molecules as in atoms; the total spin momentum $\mathbf{S}$ is given by $\sqrt{S(S+1)}$ where the total spin quantum number S is:

$$S = \Sigma S_i, \Sigma S_i - 1, \Sigma S_i - 2, \ldots, \tfrac{1}{2} \text{ or } 0. \tag{6.30}$$

The multiplicity of a molecular state is, as for atoms, $2S + 1$ and this is usually indicated as an upper prefix to the state symbol. Thus for the Π and Φ states discussed in the previous paragraph, the states would be written $^3\Pi$ or $^3\Phi$ if the individual π and δ electron spins are parallel ($S = \tfrac{1}{2} + \tfrac{1}{2} = 1$, $2S + 1 = 3$), or as $^1\Pi$ or $^1\Phi$ if the spins are paired.

When the *axial* component of a spin is required, however, it is often designated by σ for a single electron or Σ for several (corresponding to s and S for the atomic case). In this case the multiplicity is $2\Sigma + 1$.

Finally we consider the axial component of the *total* electronic angular momentum, i.e. the sum of the axial components of spin and orbital motion. In general the total momentum is strongly coupled to the axis and its axial component is more significant than the momentum itself. If we write the axial component as Ω we have simply:

$$\Omega = |\Lambda + \Sigma| \qquad (6.31)$$

but we must remember that Λ and Σ may have the same or opposite directions along the internuclear axis. Thus in the $^3\Pi$ state described above we have $\Lambda = 1$, $\Sigma = 1$, hence $\Omega = 2$ or 0. The $^1\Pi$ state has $\Lambda = 1$, $\Sigma = 0$, hence we have $\Omega = 1$ only. These states would be indicated by writing their Ω values as subscripts: $^3\Pi_2$, $^3\Pi_0$, $^1\Pi_1$.

Perhaps it will assist the student if we draw up a table (Table 6.1) showing the symbols used to designate the various sorts of angular momentum in atoms and molecules, together with their axial components, for one or more electrons.

TABLE 6.1: *Comparison of Symbols used for Electronic Angular Momenta*

	Orbital Momentum	Spin Momentum	Total Momentum
For atoms			
single electron	l (symbol s, p, d for $l = 0, 1, 2, \ldots$)	s	j
single electron (z-component)	l_z	s_z	j_z
several electrons	L (symbol S, P, D for $L = 0, 1, 2, \ldots$)	S	J
several electrons (z-component)	L_z	S_z	J_z
For molecules			
single electron	l	s	j_a (seldom used)
single electron (axial component)	λ (symbol σ, π, δ, for $\lambda = 0, 1, 2, \ldots$)	σ	ω
several electrons	L	S	J_a (seldom used)
several electrons (axial component)	Λ (symbol Σ, Π, Δ, for $\Lambda = 0, 1, 2, \ldots$)	Σ	Ω

2.4. An Example: the Spectrum of Molecular Hydrogen. Before
turning to polyatomic molecules, let us see how the above ideas may
be applied to the simplest molecule, H_2. We shall consider first
the nature of the ground state and some excited states of the molecule and how these relate to occupancy of the molecular orbitals of
Figs. 6.10, 6.12 and 6.13; then the energy of these states and, finally,
what transitions may arise between them.

The hydrogen molecule contains two electrons, one contributed
by each of the atoms. We would thus expect to find singlet and
triplet states, depending on whether the electron spins are paired or
parallel. In the *ground state* both electrons will occupy the same
lowest orbital, i.e. the $1s\sigma_g$ of Fig. 6.10 and, by Pauli's principle, they
must then form a singlet state. Both electrons are σ electrons
(since both are in a σ orbital) hence $\lambda_1 = \lambda_2 = 0$ and $\Lambda = 0$ also; the
state is thus $^1\Sigma$. We could indicate the value of Ω as a subscript
($\Omega = \Lambda + \Sigma = 0 + 0 = 0$, since $\Sigma = 0$ for singlet states) but it is more
informative to specify the symmetry (*g* or *u*) of the orbital. In this
case both electrons are in the same *g* orbital, hence the total state is
$^1\Sigma_g$.

A further subdivision of Σ states is normally made, representing
another facet of molecular symmetry. In any diatomic molecule
(whether homo- or hetero-nuclear) any plane drawn through both
nuclei is a *plane of symmetry*, i.e. electron density, shape, force
fields, etc., are quite unchanged by reflection in the plane. However, the wave-function of the electron, ψ, may either be completely
unchanged (symmetrical) or changed in sign only (antisymmetrical)
with respect to such a reflection (in either case, of course, ψ^2 is
unchanged). The former states are distinguished by a superscript
$+$ and the latter by $-$. For several reasons this division is made
for Σ states only and nearly all such states are in fact $+$. Certainly
all the states of hydrogen are symmetric.

Thus the ground state of molecular hydrogen can be written:

$$\text{ground state:}\quad (1s\sigma_g)^2 \ ^1\Sigma_g^+$$

the first part of the symbol representing the occupied orbital, the
second part the state of electrons in that orbital.

A large number of excited singlet states also exist; let us consider
some of the lower ones for which one electron only has been raised
from the ground state into some higher molecular orbital, i.e. singly
excited states.

A possible *stable* excited state thus formed (remember that the $1\sigma_u$ of Fig. 6.10 is *unstable*) is that obtained when one electron is excited into the $2s\sigma_g$ orbital, with resulting molecular configuration $(1s\sigma_g^1 2s\sigma_g^1)$. In this case both electrons are still σ electrons, hence $\Lambda = \lambda_1 + \lambda_2 = 0$, and their spins are still taken to be paired (we are discussing singlet states at present), hence $S = 0$ and the state symbol is again $^1\Sigma_0$. Both electrons are g and the overall state is also g and symmetric $(+)$:

$$\text{excited state:} \quad (1s\sigma_g 2s\sigma_g) \ ^1\Sigma_g^+.$$

On the other hand, the excited electron may be considered to arise from a $2p$ orbital. In this case (cf. Figs. 6.12 and 6.13) we have two choices, either the formation of a σ molecular orbital or of a π. The orbitals formed will not be identical with those sketched in Figs. 6.12 and 6.13 because here we are dealing with interaction between an s and a p_x or p_y orbital, on the one hand and an s and a p_z on the other, whereas the figures refer to interaction between pairs of p orbitals. However, the relevant state symbols can be derived without reference to an orbital picture. For the configuration $(1\sigma_g 2p\sigma_g)$ the state will be $^1\Sigma_0^+$ as before, but detailed calculation of the orbital shape shows this to be *odd*. Hence:

$$\text{excited state:} \quad (1s\sigma_g 2p\sigma_g) \ ^1\Sigma_u^+.$$

For $(1s\sigma_g 2p\pi_u)$, on the other hand, we now have the combination of a σ with a π electron. Hence $\Lambda = \lambda_1 + \lambda_2 = 0 + 1 = 1$ and the state is $^1\Pi_1$. Again the whole state is *odd*, so we have:

$$\text{excited state:} \quad (1s\sigma_g 2p\pi_u) \ ^1\Pi_u.$$

Thus we have arrived at three possible singlet configurations of the first excited state of hydrogen, $^1\Sigma_g^+$, $^1\Sigma_u^+$ and $^1\Pi_u$: we now need to consider the order of their energies. Firstly, remembering that $2p$ electrons reach further into space than $2s$ (cf. Fig. 5.1) it is plain that they will interact more strongly with the $1s$ electron on the other hydrogen. Since interaction is here leading to bond formation, a larger interaction forms a stronger bond, hence we would expect the orbitals formed from $2p$ electrons, the $^1\Sigma_u^+$ and $^1\Pi_u$, to be *lower* than that formed from $2s$, the $^1\Sigma_g^+$. Secondly, we can apply Hund's rule to molecules as to electrons: "for a given configuration the state of greatest multiplicity is the lowest; in states of the same multiplicity the lowest is that with the smaller L (or Λ) value."

Thus we see that the energies of the three states increase in the order

$$^1\Sigma_u^+ < {}^1\Pi_u < {}^1\Sigma_g^+.$$

Similar states are obtained by excitation to the $3s$ and $3p$ states, to the $4s$ and $4p$ states, etc. Also for $n=3, 4,\ldots$ there exists the possibility of excitation to the nd orbital. It may be shown by methods similar to those above that interaction between $1s$ and nd electrons can lead to the three configurations and state symbols in increasing energy:

$$(1s\sigma\; nd\sigma)\; {}^1\Sigma_g^+ < (1s\sigma\; nd\pi)\; {}^1\Pi_g < (1s\sigma\; nd\delta)\; {}^1\Delta_g.$$

Some of these energy levels are shown at the left of Fig. 6.14. Transitions between them can occur according to the *selection rules*:

(i) $$\Delta\Lambda = 0, \pm1. \qquad\qquad (6.32)$$

Thus transitions $\Sigma\leftrightarrow\Sigma$, $\Sigma\leftrightarrow\Pi$, $\Pi\leftrightarrow\Pi$, etc., are allowed, but $\Sigma\leftrightarrow\Delta$, for example, is not.

(ii) $$\Delta S = 0. \qquad\qquad (6.33)$$

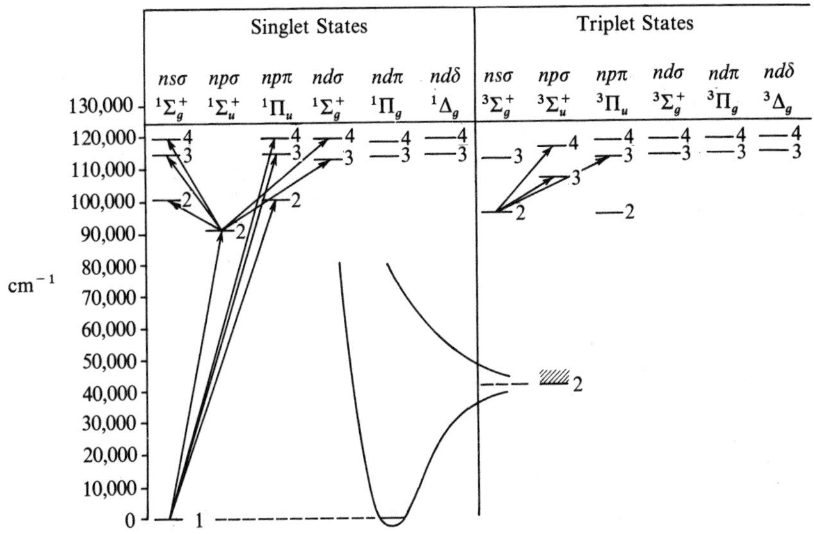

Fig. 6.14: The singlet and triplet energy levels of the hydrogen molecule; one electron only is assumed to undergo transitions, the other remaining in the 1sσ state.

For the present we are concerned only with singlet states so this rule does not arise.

(*iii*) $$\Delta\Omega = 0, \pm 1. \tag{6.34}$$

This follows directly from (*i*) and (*ii*) above.

(*iv*) There are also restrictions on symmetry changes. Σ^+ states can undergo transitions only into other Σ^+ states (or, of course, into Π states) while Σ^- go only into Σ^- (or Π). Symbolically:

$$\Sigma^+ \leftrightarrow \Sigma^+, \qquad \Sigma^- \leftrightarrow \Sigma^-, \qquad \Sigma^+ \nleftrightarrow \Sigma^-. \tag{6.35}$$

And finally:

$$g \leftrightarrow u, \qquad g \nleftrightarrow g, \qquad u \nleftrightarrow u. \tag{6.36}$$

We show a few allowed transitions from the ground state and the lowest excited state in Fig. 6.14.

Let us now consider some of the triplet states of molecular hydrogen, i.e. those states in which the electron spins are parallel and hence $S=1$. Plainly both electrons cannot now occupy the same orbital so the state of lowest energy will be either $(1s\sigma_g 2s\sigma_g)$, $(1s\sigma_g 2p\sigma_g)$ or $(1s\sigma_g 2p\pi_u)$. The first two are evidently $^3\Sigma$ states, the third is $^3\Pi$, and, following the rules outlined above, we can write down their state symbols and order of energies as:

$$(1s\sigma_g 2p\sigma_g) \,^3\Sigma_u^+ < (1s\sigma_g 2p\pi_u) \,^3\Pi_u < (1s\sigma_g 2s\sigma_g) \,^3\Sigma_g^+.$$

These energy levels are shown on the right of Fig. 6.14, together with those formed by the introduction of $3d$ and $4d$ orbitals. (The very small splitting of the levels into states with different $\Omega = \Lambda + S$ is ignored in the figure.) A few of the allowed transitions are shown from the $(1s\sigma 2s\sigma)$ state, but it should be particularly noted that, because of the selection rule $\Delta S = 0$ given in equation (6.33), transitions are not allowed between singlet and triplet states, i.e. between the two halves of Fig. 6.14.

Transitions are not shown from the lowest $^3\Sigma_u^+$ state on Fig. 6.14, i.e. the lowest triplet state. This is not because transitions are forbidden, but because the state is the continuous one shown in the upper half of Fig. 6.11, the $1s\sigma_u$. Thus in this state the molecule immediately dissociates into atoms before further transitions can occur. The energy level shown for this state in Fig. 6.14 is, in fact, the lower limit, i.e. the dissociation limit, and in fact the state extends continuously from this limit up to the top of the diagram.

Part of Fig. 6.11 is reproduced on Fig. 6.14 to underline the relationship between them.

Thus although the hydrogen spectrum will be complicated by the presence of vibrational and rotational structure on each of the transitions sketched in the figure, basically the overall pattern consists of sets of Rydberg-like line series from which the positions of the energy levels can be found. It is instructive to compare the energy level diagram for hydrogen with that given for helium in the previous chapter (Fig. 5.11, p. 175). The singlet and triplet S, P and D states of helium become the singlet and triplet Σ, Π and Δ states of hydrogen, with the added refinement of g and u configurations. Essentially, however, the energy level patterns are similar and the spectra will also be similar, except in so far as the helium spectrum shows no vibrational or rotational structure.

3. Electronic Spectra of Polyatomic Molecules

We have seen in Chapter 3, Section 7 that the vibrational frequencies of a particular atomic grouping within a molecule, e.g. CH_3, $C{=}O$, $C{=}C$, etc., are usually fairly insensitive to the nature of the rest of the molecule. Other bond properties, such as length or dissociation energy, are also largely independent of the surrounding atoms in a molecule. Since all these properties depend, in the final analysis, on the electronic structure of the bond, it is plain that we may, at least as an approximation, discuss the structure, and hence the spectrum, of each bond in isolation. Bonds for which this approximation is adequate are usually said to have "localized" molecular orbitals, i.e. orbitals embracing a pair of nuclei only; other molecules, for which this approximation is invalid, have non-localized orbitals and are often called "conjugated". We shall meet some examples of this latter class shortly.

When each bond may be considered in isolation, it is evident that the complete electronic spectrum of a molecule is the sum of the spectra from each bond. The result will plainly be very complex, but a great deal of information about the molecule is contained within it. Thus if some band series can be recognized for a particular bond we immediately know the vibrational frequency of that bond and probably a good estimate of its dissociation energy also. If the rotational structure is resolved, then we have the moment of inertia (from the line spacing) and hence information about the shape and size of the molecule.

Such detailed information is usually obtainable only for molecules studied in the gas phase: in pure liquids, or in solution, molecular rotation is hindered and no rotational structure will be observed. The blurring of the rotational structure often masks the vibrational line series also, and the electronic spectrum of a liquid is usually rather broad and characterless. However, as we shall discuss shortly, it may still be highly characteristic of a particular molecular grouping both in its frequency and its intensity.

Confining our attention, for the moment, to gas-phase spectra, we have already remarked that one of the more important advantages of electronic spectroscopy is that the vibrations, rotations, dissociation energies and structures of molecules may be investigated in their *excited states*, even though a particular molecule may exist in such a state for not much longer than the time it takes to complete a few rotations. We have not the space to discuss this topic in detail, but one aspect is especially interesting—the fact that electronic excitation often leads to a change in shape of the molecule. That this happens can be seen by studying rotational fine structure in the spectra; here we briefly discuss the theoretical basis for its occurrence.

3.1. Change of Shape on Excitation. In Fig. 6.15 we show the orbital picture of a hydride H_2A, where A is any polyvalent atom, both in its bent configuration (*a*), with a bond angle of 90°, and in a linear form (*b*), bond angle 180°. We have seen in Chapter 5, Section 1, that the p orbitals of an atom are at right angles to each other, so we can readily imagine the rectangular molecule to be formed by interaction of two of these p orbitals with hydrogen $1s$ orbitals, leaving the third p orbital unaffected; the latter is called a "non-bonding" orbital since it plays no part in bonding A and H together, and in general it will have a higher energy than bonding orbitals. In Fig. 6.15(*a*) we label the non-bonding orbital (which is out of the plane of the paper) N_1, the two bonding molecular orbitals formed from the p and $1s$ atomic orbitals as a_1 and a_2 and the unused (and hence non-bonding) atomic orbital s of A simply as s.

In the linear molecule a new principle must be introduced—that of orbital hybridization. For this, atom A is supposed to mix its s orbital with one of its p orbitals and, from these two orbitals, to form two new orbitals—hybrids— which, it may be shown theoretically, point at 180° to each other. These sp hybrids form rather stronger bonds to other atoms than separate s and p orbitals, so it is energetically favourable, in certain cases, for the atom to "prepare" hybrid orbitals at the moment of bond formation. In this configuration (Fig. 6.15(b)), there are now two non-bonding p orbitals, labelled N_1 and N_2, and two bonding orbitals formed by overlap between sp hybrids and hydrogen $1s$, called b_1 and b_2.

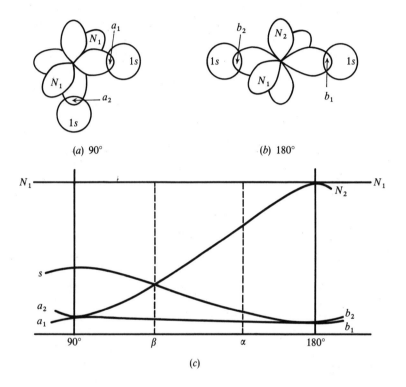

Fig. 6.15: The orbital pictures for an AH_2 molecule where the AH bonds are (a) at 90°, and (b) at 180°. In (c) is shown qualitatively the change in energy of the various orbitals as the bond angle changes from 90 to 180°. (Adapted, with the kind permission of the author, from A. D. Walsh, J. Chem. Soc., 1953, page 2262.)

Now, remembering that a non-bonding orbital is higher in energy than a bonding, and that an sp-bonding orbital is stronger (hence *lower* in energy because the molecule is more stable) than a p-bonding, we can plot the qualitative energy changes for a smooth transition from 90 to 180° bonding.   This we do in Fig. 6.15(c), which is constructed as follows:

(i) The non-bonding orbital N_1 remains unchanged throughout, hence its energy is constant;

(ii) The bonding orbital a_1 passes over into the stronger orbital b_1 hence its energy decreases;

(iii) The bonding orbital a_2 becomes the non-bonding N_2, thus increasing in energy; N_1 and N_2 are identical in energy at 180°;

(iv) The bonding orbitals b_1 and b_2 are formed by absorption of the non-bonding s into a_1;

(v) If we increase the bond angle beyond 180° (or decrease it below 90°) the reverse changes begin to take place, so 180° and 90° represent maxima and minima on the energy curves as shown.

Now let us see the relevance of this to molecular shapes.   Consider first the molecule BeH_2, beryllium hydride.   Beryllium, we have seen in Chapter 5, Section 3.1, has the electronic ground state configuration $1s^2 2s^2$, i.e. it has two outer electrons with which to form bonds, the two $1s$ electrons being too firmly held by the nucleus to take part in bonding.   Each hydrogen atom contributes a further electron, so the BeH_2 molecule must dispose of four electrons into molecular orbitals, with, according to Pauli, a maximum of two electrons per orbital.   The most stable state, as can be seen from Fig. 6.15(c) will be for two electrons to go into b_1 and two into b_2, thus producing a *linear* molecule.

When the molecule is excited electronically, the next available orbital to contain the excited electron is N_2 (or N_1), but with a configuration $b_1^2 b_2^1 N_2^1$ it is evident from the figure that the most stable state will be at a bond angle, α, somewhere between 90 and 180°—the increase in the energies of b_1 and b_2 being more than compensated for by the decrease in N_2 until equilibrium occurs at an angle α.   Thus we see that the excited state is *bent*.   If the electron is so excited as to be ionized completely, leaving the ion BeH_2^+, the three remaining electrons will all be in the b_1 and b_2 orbitals and hence the most stable configuration will again be *linear*.

Now consider the case of water, H_2O.   The oxygen atom has an outer electron configuration $2s^2 2p^4$, and so has six electrons to dispose into molecular orbitals.   As before each hydrogen contributes one, so water is formed by placing a total of four pairs of electrons into four molecular orbitals. The lowest energy state at which this can be done is shown by the angle β in the figure, which is some angle between 90 and 180° (and is observed experimentally to be about 104°). Thus water is bent in the ground state, with a configuration which may be written $a_1^2 a_2^2 s^2 N_1^2$, since the angle is not far removed from 90°. During excitation one of the N_1 electrons will undergo transitions since these, being of highest energy, are most easily removed. However, since the energy of N_1 does not change with angle, the angle of the remaining $a_1^2 a_2^2 s^2 N_1^1$ state will not change during the transition.

These arguments may be readily extended to other triatomic molecules or to larger polyatomic molecules, although the energy diagram corresponding to Fig. 6.15(c) is more complicated since more orbitals are involved. The results show, however, and experiment confirms, that linear molecules such as CO_2 and $HC\equiv CH$ become bent on excitation, the latter taking up a planar zig-zag conformation.

3.2. Chemical Analysis by Electronic Spectroscopy. Although rotational and sometimes vibrational fine structure do not appear in the liquid or solid state, both the position and intensity of the rather broad absorption due to an electronic transition is very characteristic of the molecular group involved. In this branch of spectroscopy the position of an absorption is almost invariably given as the *wave-length* at the point of maximum absorption, $\lambda_{\max}$, quoted either in Ångstrom units (1 Å = 10^{-8} cm) or in millimicrons (1 mμ = 10^{-7} cm), the latter being more usual. It should be particularly noted that a *large* energy change, corresponding to a high frequency or wavenumber, is represented by a *small* wave-length. For practical reasons the electronic spectrum is divided into three regions: (*i*) the visible region, between 4000 and 7500 Å (400–750 mμ or 25,000–13,300 cm^{-1}), (*ii*) the near ultra-violet region, between 2000 and 4000 Å (200–400 mμ, or 50,000–25,000 cm^{-1}), and (*iii*) the far (or vacuum) ultra-violet, below 2000 Å (below 200 mμ or above 50,000 cm^{-1}). The latter is so called because absorption by atmospheric oxygen is considerable in this region and spectra can only be obtained if the whole spectrometer is carefully evacuated. Thus commercial instruments extend only down to about 185 mμ and absorptions below this range are little used for routine chemical purposes.

The *intensity* of an electronic absorption is given by the simple equation:

$$\varepsilon = \frac{1}{cl} \log_{10} \frac{I_0}{I}, \qquad (6.37)$$

where c and l are the concentration and path length of the sample, I_0 is the intensity of light of wave-length $\lambda_{max.}$ falling on the sample, and I is the intensity transmitted by the sample. ε is the extinction coefficient and ranges from some 5×10^5 for the strongest bands to 1 or less for very weak absorptions.

Electrons in the vast majority of molecules fall into one of the three classes: σ-electrons, π-electrons, and non-bonding electrons (called n-electrons). The first two classes were discussed in Section 2.2 of this chapter and the third, which plays no part in the bonding of atoms into molecules, was mentioned briefly in Section 3.1. In chemical terms a single bond between atoms, such as C—C, C—H, O—H, etc., contains only σ-electrons, a multiple bond, C=C, C≡C, C=N, etc., contains π-electrons in addition, while atoms to the right of carbon in the periodic table, notably nitrogen, oxygen and the halogens, possess n-electrons. In general the σ-electrons are most firmly bound to the nuclei and hence require a great deal of energy to undergo transitions, while the π- and n-electrons require less energy, the n-electrons usually (but not invariably) requiring less than the π. Thus, in an obvious notation, $\sigma \rightarrow \sigma^*$ transitions fall in the vacuum ultra-violet, $\pi \rightarrow \pi^*$ and $n \rightarrow \sigma^*$ appear near the borderline of the near and far ultra-violet, and $n \rightarrow \pi^*$ come well into the near ultra-violet and visible regions. These generalizations are indicated schematically on Fig. 6.16, which also shows the relationship between the millimicron and wavenumber scales.

Saturated hydrocarbon molecules, then, which can only undergo $\sigma \rightarrow \sigma^*$ transitions, do not give rise to spectra with any analytic interest since they fall outside the generally available range; examples are the $\sigma \rightarrow \sigma^*$ transitions of methane CH_4, and ethane C_2H_6 which are at 122 and 135 mμ respectively.

The insertion of a group containing n-electrons, e.g. the NH_2 group, allows the possibility of $n \rightarrow \sigma^*$ transitions in addition and also tends to increase the wave-length of the $\sigma \rightarrow \sigma^*$ absorption; e.g. CH_3NH_2: $\sigma \rightarrow \sigma^*$ 170 mμ, $n \rightarrow \sigma^*$ 213 mμ. It is unsaturated molecules, i.e. molecules containing multiple bonds, which give rise to the most varied and interesting spectra, however. We cannot here discuss the large mass of data in any detail but must be content to indicate a few of the more important generalizations. More detail is to be found in the books by Rao and by Phillips listed in the bibliography at the end of this chapter.

Consider, first, *isolated* multiple bonds within a molecule; the most important factor determining the position of the absorption

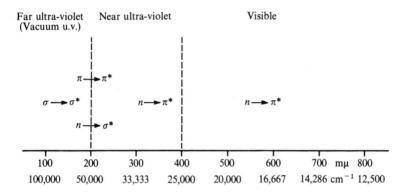

Fig. 6.16: *The regions of the electronic spectrum and the type of transition which occurs in each.*

maxima is, of course, the nature of the atoms which are multiply-bonded. From the following table we see that the $\pi \rightarrow \pi^*$ transitions are relatively insensitive to those atoms while the $n \rightarrow \pi^*$ transitions vary widely:

	$\pi \rightarrow \pi^*$ (strong) (mμ)	$n \rightarrow \pi^*$ (weak) (mμ)
$>$C=C$<$	170	—
—C≡C—	170	—
$>$C=O	166	280
$>$C=N$\backslash$	190	300
$/$N=N$/$	?	350
$>$C=S	?	500

This behaviour is very reasonable since the *n*-electrons play no part in the bonding and control of them is retained by the atom (O, N or S) contributing them. The above data is approximate only since different substituents on the A=B group produce slight variations in the wave-length of the $n \rightarrow \pi^*$ transition. Thus, considering ketones alone, λ_{max} varies from 272 mμ for CH_3COCH_3 to 290 mμ

for cyclohexanone, and even higher if halogen substituents are included.   From the mass of empirical data already assembled, a great deal of information about the substituents to a particular group is obtainable from the electronic spectrum.

More pronounced changes occur, however, when two or more multiple bonds are *conjugated* in the molecule, i.e. when structures having alternate single and multiple bonds arise, for example —C=C—C=C— or —C=C—C=O.   In this case the $\pi \rightarrow \pi^*$ and $n \rightarrow \pi^*$ transitions both increase considerably in wave-length and intensity, the increase being greater the more conjugate linkages there are.   As a simple example we have the following approximate data for $\pi \rightarrow \pi^*$ transitions in carbon–carbon bonds:

	$\lambda_{max.}$ (mμ)	ε
—C=C—	170	16,000
—C=C—C=C—	220	21,000
—C=C—C=C—C=C—	260	35,000

while for oxygen-containing molecules we have both $\pi \rightarrow \pi^*$ and $n \rightarrow \pi^*$ transitions:

	$\pi \rightarrow \pi^*$ (strong) (mμ)	$n \rightarrow \pi^*$ (weak) (mμ)
—C=O	166	280
—C=C—C=O	240	320
—C=C—C=C—C=O	270	350
O=⟨ ⟩=O	245	435

Thus we see that conjugation immediately brings the very intense $\pi \rightarrow \pi^*$ transition into the easily available region of ultra-violet spectrometers.   For this reason these techniques are particularly well-adapted to the study of conjugated and aromatic systems.

In the last example given above we see that the $n \rightarrow \pi^*$ absorption of *p*-benzoquinone, at 435 mμ, has shifted into the blue region of the visible spectrum.   When the substance is seen in the ordinary way by reflected light the complementary colour—yellow—is observed. Colour in large organic molecules is invariably due to the existence of considerable conjugation raising the transition wave-length into the visible region—a fact on which the chemistry of dyestuffs is based.

3.3. The Re-Emission of Energy by an Excited Molecule. After a molecule has undergone an electronic transition into an excited state there are several processes by which its excess energy may be lost; we discuss some of these briefly below.

(*i*) Dissociation. The excited molecule breaks into two fragments. This was discussed in some detail for the particular case of a diatomic molecule dissociating into atoms in Section 1.4. No spectroscopic phenomena, beyond the initial absorption spectrum,

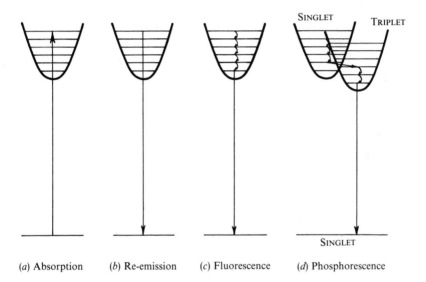

SINGLET   TRIPLET

SINGLET

(*a*) Absorption   (*b*) Re-emission   (*c*) Fluorescence   (*d*) Phosphorescence

Fig. 6.17: Showing the various ways in which an electronically excited molecule can lose energy.

are observed unless the fragments radiate energy by one of the processes mentioned below.

(*ii*) Re-Emission. If the absorption process takes place as shown schematically in Fig. 6.17(*a*), then the re-emission is just the reverse of this, as in (*b*) of the figure. The radiation emitted, which may be collected and displayed as an emission spectrum, is identical in frequency with that absorbed.

(*iii*) Fluorescence. If, as in Fig. 6.17(*a*), the molecule is in a high *vibrational* state after electronic excitation, then excess vibrational energy may be lost by intermolecular collisions; this is illustrated

16

in (c) of the figure.   The vibrational energy is converted to kinetic energy and appears as heat in the sample; such transfer between energy levels is referred to as "radiationless".   When the excited molecule has reached a lower vibrational state (e.g. $v' = 0$), it may then emit radiation and revert to the ground state; the radiation emitted, called the *fluorescence spectrum*, is normally of lower frequency than that of the initial absorption, but under certain conditions it may be of higher frequency.   The time between initial absorption and return to ground state is very small, of the order of 10^{-8} sec.

(iv) Phosphorescence.   This can occur when two excited states of different total spin have comparable energies.   Thus in (d) of Fig. 6.17, we imagine the ground state and one of the excited states to be singlets (i.e. $S = 0$), while the neighbouring excited state is a triplet ($S = 1$).   Although the rule $\Delta S = 0$ forbids *spectroscopic* transitions between singlet and triplet states, there is no prohibition if the transfer between the excited states occurs *kinetically*, i.e. through radiationless transitions induced by collisions.   Such transfer, however, can only occur close to the cross-over point of the two potential curves (cf. Section 1.7), and once the molecule has arrived in the triplet state and undergone some loss of vibrational energy in that state, it cannot return to the excited singlet state.   It will, therefore, eventually reach the $v' = 0$ level of the *triplet* state. Now although a transition from here to the ground state is spectroscopically forbidden, it *may* take place but much more slowly than an allowed electronic transition.   Thus it is that a phosphorescent material will continue to emit radiation seconds, minutes or even hours after the initial absorption.   The phosphorescence spectrum, as a rule, consists of frequencies lower than that absorbed.

Considerable confusion often occurs between the Raman effect, discussed in Chapter 4, and the phenomena of fluorescence or phosphorescence.   The main points of difference are as follows:

(i) In fluorescence and phosphorescence, radiation must be absorbed by the molecule and an excited electronic state formed; in Raman spectroscopy energy is merely transferred from radiation to molecule, or vice versa, but no excited electronic state is formed.

(ii) The exciting radiation for fluorescence or phosphorescence must be just that equivalent to the energy difference between electronic states; the exciting radiation for Raman spectroscopy can be of any frequency *except* that which would induce electronic transitions; in the latter case absorption would occur rather than scattering.

4. Techniques and Instrumentation

The simple techniques of electronic spectroscopy are familiar to every schoolboy studying physics—a glass prism, some sort of telescope, a bunsen burner and a pinch of common salt are sufficient apparatus for observing part of the emission spectrum of sodium. And in fact a great deal of rapid and precise analytical work, both qualitative and quantitative, is carried out using flame spectrophotometers not very much more sophisticated in construction than this, except that a photomultiplier or photographic plate is used instead of the rather inaccurate human eye. However, for high-resolution work or for absorption studies, the practical requirements are more stringent.

The choice of a suitable source was formerly one of the main difficulties. The prime requirements of a source are that it should be *continuous* over the region of interest (i.e. there must be no wavelengths at which it does not emit) and it should be as *even* as possible (i.e. there must be no intense emission lines). In the visible region and just into the near ultra-violet—say between 350 and 800 mμ—an ordinary tungsten filament lamp is quite suitable. Below this a hydrogen discharge lamp proves adequate, down to about 190 mμ while below this again discharge lamps containing rare gases, such as xenon, must be used. Thus we see that, in contrast to the other forms of spectroscopy discussed in previous chapters, no one source is suitable throughout the region.

Transparent materials for windows and sample cells present no problem, at least in the visible and near ultra-violet regions, since good quality glass or quartz transmit down to 200 mμ or better. Below this region alkali fluorides, such as lithium fluoride or calcium fluoride must be used; these are transparent down to about 100 mμ. Prisms, if used, can be made of the same materials. Modern high-resolution instruments, however, employ a reflection grating rather than a prism since the former gives better dispersion and so allows more precise wave-length selection and measurement.

The detector for visible and ultra-violet studies is either a photographic plate or a photomultiplier tube. The chief disadvantage of the photographic method is that the resolving power is limited by the graininess of the image; on the other hand there is no other detector which can record the complete spectrum simultaneously in a small fraction of a second. When studying short-lived species, such as free radicals, it would be quite impossible to scan the complete spectrum using a photomultiplier. Also, at the other end of the time-scale the photographic plate is an efficient integrator of very weak signals—exposure times can be extended to many hours or even days to record a weak emission or absorption. For most routine purposes, however, where the spectrum of a stable material is to be recorded in a time of several minutes, a photomultiplier detector coupled to an amplifier and paper chart recorder is the most flexible and useful combination.

Bibliography

BARROW, G. M.: *Introduction to Molecular Spectroscopy*, McGraw-Hill, 1962

BERL, W. G. (Ed.): *Physical Methods in Chemical Analysis*, Vol. I, 2nd edn., Academic Press, 1960

HERZBERG, G.: *Molecular Spectra and Molecular Structure;* Vol. 1, *Spectra of Diatomic Molecules*, 2nd edn., Van Nostrand, 1950

KING, G. W.: *Spectroscopy and Molecular Structure*, Holt, Rinehart and Winston Inc., 1964

NACHOD, F. C. and W. D. PHILLIPS: *Determination of Organic Structures by Physical Methods*, Vol. 2, Academic Press, 1962

RAO, C. N. R.: *Ultra-violet and Visible Spectroscopy*, Butterworths, 1961

WALKER, S. and H. STRAW: *Spectroscopy*, Vol. II, Chapman & Hall, 1962

CHAPTER 7

SPIN RESONANCE SPECTROSCOPY

We have seen in earlier chapters that all electrons and some nuclei possess a property conveniently called "spin". Electronic spin was introduced in Chapter 5 to account for the way in which electrons group themselves about a nucleus to form atoms and we found that the spin also accounted for some fine structure, such as the doublet nature of the sodium D line, in atomic spectra. Equally, in Section 6 of Chapter 5 it was necessary to invoke a nuclear spin to account for very tiny effects, called hyperfine structure, observed in the spectra of some atoms.

In this chapter we shall consider these spins in rather more detail and discuss the sort of spectra they can give rise to directly rather than their influence on other types of spectra. After an introduction discussing the interaction of spin with an external magnetic field we shall consider in some detail the spectra of particles with a spin of $\frac{1}{2}$ (i.e. electrons, and some nuclei such as hydrogen, fluorine or phosphorus), then a brief discussion of some other nuclei whose spins are greater than $\frac{1}{2}$, and finally a few words on the techniques involved in producing electronic and nuclear spin spectra.

1. Spin and an Applied Field

1.1. The Nature of Spinning Particles. We have seen that all electrons have a spin of $\frac{1}{2}$, i.e. they have an angular momentum of $\sqrt{\frac{1}{2}(\frac{1}{2}+1)}(h/2\pi) = \sqrt{3}/2$ a.m. units. Many nuclei also possess spin although the angular momentum concerned varies from nucleus to nucleus.

The simplest nucleus is that of the hydrogen atom, which consists of one particle only, the *proton*. The protonic mass $(1.66 \times 10^{-24}$ g) and charge $(+4.80 \times 10^{-10}$ e.s.u.) are taken as the units of atomic mass and charge respectively; the charge is, of course, equal in magnitude but opposite in sign to the electronic charge. The proton also has a spin of $\frac{1}{2}$.

Another particle which is a constituent of all nuclei (apart from the hydrogen nucleus) is the *neutron*; this has unit mass (i.e. a mass equal to that of the proton), no charge and, again, a spin of $\frac{1}{2}$.

Thus if a particular nucleus is composed of p protons and n neutrons its total mass is $p+n$ (ignoring the small mass defects associated with nuclear binding energy), its total charge is $+p$ and its total spin will be a vector combination of $p+n$ spins each of magnitude $\frac{1}{2}$. The atomic mass is usually specified for each nucleus by writing it as a prefix to the nuclear symbol, e.g. ^{12}C indicates the nucleus of carbon having a mass of 12. Since the atomic charge is six for this nucleus we know immediately that the nucleus must contain six protons and six neutrons to make up a mass of 12. The nucleus ^{13}C (an *isotope* of carbon) has six protons and seven neutrons.

Each nuclear isotope, being composed of a different number of protons and neutrons will have its own total spin value. Unfortunately, the laws governing the vector addition of nuclear spins are not yet known so the spin of a particular nucleus cannot be predicted in general. However, observed spins can be rationalized and some empirical rules have been formulated.

Thus the spin of the hydrogen nucleus (^{1}H) is $\frac{1}{2}$ since it consists of one proton only; deuterium, an isotope of hydrogen containing one proton and one neutron (i.e. ^{2}H) might have a spin of 1 or zero depending on whether the proton and neutron spins are parallel or opposed: it is observed to be 1. The helium nucleus, containing two protons and two neutrons (^{4}He) has zero spin, and from these and other observations stem the following rules:

(*i*) Nuclei with both p and n even (hence charge *and* mass even) have zero spin (e.g. ^{4}He, ^{12}C, ^{16}O, etc.).
(*ii*) Nuclei with both p and n odd (hence charge *odd* but mass $= p+n$, *even*), have integral spin (e.g. ^{2}H, ^{14}N (spin $=1$), ^{10}B (spin $= 3$), etc.).
(*iii*) Nuclei with odd mass have half-integral spins (e.g. ^{1}H, ^{15}N (spin $=\frac{1}{2}$), ^{17}O (spin $=\frac{5}{2}$), etc.).

The spin of a nucleus is usually given the symbol I, called the *spin quantum number*. Quantum mechanics shows that the angular momentum of a nucleus is given by the expression:

angular momentum $\mathbf{I} = \sqrt{I(I+1)}(h/2\pi) = \sqrt{I(I+1)}$ a.m. units

$$(7.1)$$

where I takes, for each nucleus, *one* of the values $0, \frac{1}{2}, 1, \frac{3}{2}, \ldots$. We can conveniently include the spin quantum number and angular momentum of an *electron* in equation (7.1) if we agree to label its spin quantum number I (instead of s as in Chapter 5), and remember that I can be $\frac{1}{2}$ only for an electron. Thus equation (7.1) represents the angular momentum of nuclei and of electrons once the appropriate value of I is inserted.

We may note here that many texts use a simpler form of equation (7.1), viz.

$$\mathbf{I} = I\frac{h}{2\pi} = I \text{ a.m. units.}$$

This equation is not, however, strictly correct from a quantum mechanical point of view and we shall use the more rigorous equation (7.1) throughout this chapter.

By now the reader will be familiar with the idea that the angular momentum vector $\mathbf{I}$ cannot point in any arbitrary direction, but can point only so that its components along a particular reference direction are either all integral (if I is integral) or all half-integral (if I is half-integral). Thus we can have components along a particular direction z, defined by:

$$I_z = I, I-1, \ldots, 0, \ldots, -(I-1), -I \quad \text{(for } I \text{ integral)}$$

or

$$I_z = I, I-1, \ldots, \tfrac{1}{2}, -\tfrac{1}{2}, \ldots, -I \quad \text{(for } I \text{ half-integral)} \quad (7.2)$$

giving $2I+1$ components in each case. These components are normally degenerate—i.e. they all have the same energy—but the degeneracy may be lifted and $2I+1$ different energy levels result if an external magnetic field is applied to define the reference direction. We shall now consider the effect of such a field.

1.2. Interaction between Spin and a Magnetic Field. In general, a charged particle spinning about an axis constitutes a circular electric current which in turn produces a magnetic dipole. In other words the spinning particle behaves as a tiny bar magnet placed along the spin axis. The size of the dipole (i.e. the strength of the magnet) for a point charge can be shown to be:

$$\boldsymbol{\mu} = \frac{q}{2mc}\mathbf{I} = \frac{q\sqrt{I(I+1)}}{2mc}\frac{h}{2\pi}, \quad (7.3)$$

where q and m are the charge and mass of the particle and c is the velocity of light. When we remove the fiction that electrons and nuclei are *point* charges, equation (7.3) becomes modified by the inclusion of a numerical factor G:

$$\mu = \frac{Gq\sqrt{I(I+1)}}{2mc}\frac{h}{2\pi}. \qquad (7.4)$$

For electrons we have seen (cf. Chapter 5, Section 5) that G is given the symbol g and called the Landé splitting factor; its value depends on the quantum state of the electron and may be calculated from the L, S and J quantum numbers (cf. equation (5.28)). Nuclear G-factors, on the other hand, may not be calculated in advance and are obtainable only experimentally.

For electrons, equation (7.4) is usually written

$$\mu = -g\beta\sqrt{I(I+1)}, \qquad (7.5)$$

where we have expressed the set of constants $eh/4\pi mc$ as a (positive) constant β called the *Bohr magneton*. The electronic charge and mass are respectively, $-4\cdot80 \times 10^{-10}$ e.s.u. and $9\cdot11 \times 10^{-28}$ g, and we can calculate $\beta = 9\cdot273 \times 10^{-21}$ erg. gauss^{-1}.

Nuclear dipoles, on the other hand, are conveniently expressed in terms of a *nuclear magneton* β_N, which is defined in terms of the mass and charge of the *proton*:

$$\beta_N = \frac{eh}{4m_p\pi c} = 5\cdot050 \times 10^{-24} \text{ erg. gauss}^{-1}.$$

Thus for a nucleus of mass M and charge pe (where p is the number of protons) we would write:

$$\mu = \frac{Gpe}{2Mc}\sqrt{I(I+1)}\frac{h}{2\pi} = \frac{Gm_p p}{M}\beta_N\sqrt{I(I+1)}$$

$$= g\beta_N\sqrt{I(I+1)}, \qquad (7.6)$$

where we have collected the parameters $Gm_p p/M$, in which m_p is the protonic mass, into a factor g which is characteristic of each nucleus. This factor has values up to about six and is positive for nearly all known nuclei.

Thus the analogous equations (7.5 and 7.6) define the equivalent spin dipole for any spinning particle. The dipole will plainly have components along a reference direction governed by the I_z values:

$$\left.\begin{array}{l} \mu_z = -g\beta I_z \text{ (for electrons)} \\ \mu_z = g\beta_N I_z \text{ (for nuclei)} \end{array}\right\} \tag{7.7}$$

where the I_z are given by equation (7.2) for a particular particle, and the dipole will interact to different extents, depending on its magnitude, with a magnetic field. The situation for a *nucleus* with $I = 1$ is shown in Fig. 7.1. The angular momentum of the particle (equation (7.1)) is:

$$\mathbf{I} = \sqrt{1 \times 2} = \sqrt{2} \text{ a.m. units}$$

and if we consider a semicircle with this radius, it is plain that the

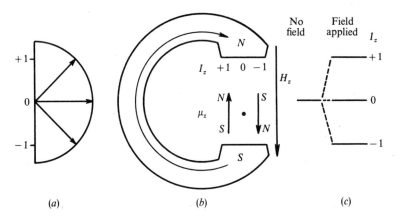

Fig. 7.1: Showing (a) *the three orientations of the spin of a nucleus with spin quantum number $I=1$; (b) the resulting magnetic dipole, μ_z, oriented in an applied magnetic field H_z; and (c) the three energy levels allowed to the nucleus.*

vector arrow corresponding to $\mathbf{I}$ can point so as to have z-components of $+1$, 0 or -1 (the z-direction is counted positive towards the top of the paper). This is shown in part (a) of the figure.

Equation (7.7) shows that μ_z and I_z have the same sign (i.e. point in the same direction) for the many nuclei which have positive g-values. The three μ_z-values for this system are represented by arrows in part (b) of the figure where, conventionally, the magnetic dipole is shown by an arrow drawn *inside* the nuclear magnet with its head pointing to the N pole. When the same convention is followed with the applied field, i.e. $S \rightarrow N$ inside the magnet, the magnetic lines of force *outside* the magnet must be represented by an arrow from N to S as shown. Thus we see that the state $I_z = +1$ represents a nuclear dipole *opposed* to the magnetic field (i.e. of high energy) while $I_z = -1$ is in the same direction as the applied field and is, therefore, of low energy. The state $I_z = 0$ has no net dipole along the field direction and is therefore unchanged in energy whether the field is applied or not. This is shown in Fig. 7.1(*c*).

Of course, if the nuclear *g*-factor is negative, μ_z has a sign opposite from I_z and the order of labelling the energy levels of Fig. 7.1 will be reversed. Similarly, of the two energy levels allowed to an electron in a magnetic field, the lower will be associated with $I_z = +\frac{1}{2}$, the upper with $I_z = -\frac{1}{2}$.

The extent of interaction between a magnetic dipole and a field of strength H_z applied along the z axis is equal to the product of the two:

$$\text{interaction} = \mu_z H_z.$$

Thus the separation between neighbouring energy levels (where I_z differs by unity) is:

$$\Delta E = [E_{I_z} - E_{(I_z - 1)}] = |g\beta_N I_z H_z - g\beta_N (I_z - 1)H_z|$$

$$= |g\beta_N H_z| \text{ ergs} \quad \text{(when } H_z \text{ is expressed in gauss)}.$$

Thus in c/sec:

$$\frac{\Delta E}{h} = \left|\frac{g\beta_N H_z}{h}\right| \text{ c/sec} \qquad (7.8)$$

(where the modulus sign, $|\ldots|$, indicates that *positive* differences only should be considered). Here, then, is the basis for a spectroscopic technique: a transition of electron or nuclear spins between energy levels (loosely referred to as "a change of spin") may be associated with the emission or absorption of energy in the form of radiation at the frequency of equation (7.8). Further, since the frequency is proportional to the applied field we can arrange, in principle, to study spin spectra in any region of the electromagnetic spectrum, merely by choosing an appropriate field. However, for practical reasons, the fields used are normally of the order of 15,000 gauss for nuclei and 3000 gauss for electrons. Let us calculate the approximate frequency to be expected under these circumstances.

For nuclei: We have already $\beta_N = 5 \cdot 05 \times 10^{-24}$ erg. gauss^{-1}, and if we choose $H_z = 15,000$ gauss and the g-factor corresponding to that of hydrogen, $g = 5·585$, we calculate

$$\frac{\Delta E}{h} = \frac{5·585 \times 5·05 \times 10^{-24} \times 1·5 \times 10^4}{6·63 \times 10^{-27}} \approx 64 \times 10^6 \text{ c/sec.} \quad (7.9)$$

Thus we see the appropriate frequency for protons, 64 megacycles/second (Mc/sec), falls in the region of short-wave radio frequencies. All other nuclei (except tritium) have smaller g-factors and their spectra fall between 10 and 60 Mc/sec for the same applied field.

For electrons: Here $\beta = 9·273 \times 10^{-21}$ erg. gauss^{-1}, and let us assume $g = 2$ and $H_z = 3000$ gauss. Then

$$\frac{\Delta E}{h} = \frac{2 \times 9·273 \times 10^{-21} \times 3 \times 10^3}{6·63 \times 10^{-27}} = 8,400 \times 10^6 \text{ c/sec.} \quad (7.10)$$

Thus electron spin spectra fall at a considerably higher frequency, which is on the long wave-length edge of the microwave region. Because of this difference, techniques of nuclear and electronic spin spectroscopy differ considerably, as we shall see later, although in principle they are concerned with very similar phenomena.

1.3. Population of Energy Levels. When first confronted with nuclear and electron spin spectroscopy the student (who has experimented earlier with bar magnets in the earth's field) usually asks: why don't the nuclear (or electronic) magnetic moments immediately line themselves up in an applied field so that they all occupy the lowest energy state?

There are several facets to this question and its answer. Firstly, if we take "immediately" to refer to a period of some seconds, then spin magnetic moments *do* immediately orientate themselves in a magnetic field, although they do *not* all occupy the lowest available energy state. This is a simple consequence of thermal motion and the Boltzmann distribution. We have seen that spin energy levels are split in an applied field, and their energy separation (equation (7.8)) is ΔE ergs. Let us confine our attention to particles with spin, $\frac{1}{2}$ (and hence just two energy levels) for simplicity—our remarks, however, are easily extended to cover the general case. Classical theory states that at a temperature $T°$ Abs. the ratio of the populations of such levels will be given by

$$\frac{N_{\text{upper}}}{N_{\text{lower}}} = \exp\left(-\frac{\Delta E}{T}\right) \quad (7.11)$$

where k is the Boltzmann constant. Thus at all temperatures above absolute zero the upper level will always be populated to some extent, although for large ΔE the population may be insignificant. In the case of nuclear and electron spins, however, ΔE is extremely small:

$$\Delta E \text{ nuclei} \approx 5 \times 10^{-19} \text{ erg in 15,000 gauss field}$$

$$\Delta E \text{ electrons} \approx 5 \times 10^{-17} \text{ erg in 3000 gauss field}$$

and, since $K = 1\cdot38 \times 10^{-16}$ erg. degree^{-1}, we have at room temperature ($T = 300°$ Abs.)

$$\frac{N_{\text{upper}}}{N_{\text{lower}}} \approx \exp\left(-\frac{5 \times 10^{-19}}{4\cdot2 \times 10^{-14}}\right)$$

$$\approx \exp(-1\cdot2 \times 10^{-5}) \approx 1 - 1\cdot2 \times 10^{-5} \quad \text{for nuclei}$$

or

$$\frac{N_{\text{upper}}}{N_{\text{lower}}} \approx \exp\left(-\frac{5 \times 10^{-17}}{4\cdot2 \times 10^{-14}}\right)$$

$$\approx \exp(-1\cdot2 \times 10^{-3}) \approx 1 - 1\cdot2 \times 10^{-3} \quad \text{for electrons.}$$

In both cases the ratio is very nearly equal to unity and we see that the spins are almost equally distributed between the two (or, in general, $2I + 1$) energy levels.

We need to discuss now the nature of the interaction between radiation and the particle spins which can give rise to transitions between these levels.

1.4. The Larmor Precession. We have seen (equation (7.6)) that the dipole moment of a spinning nucleus is

$$\mu = g\beta_N\sqrt{I(I+1)}$$

and that, according to quantal laws, the vector represented by μ can be oriented only so that its components are integral or half-integral in a reference direction. The corollary to this is that, since $\sqrt{I(I+1)}$ cannot be integral or half-integral, if I is integral or half-integral, the vector arrow can *never* be exactly in the field direction. We see one example of this in Fig. 7.1(a) and another for a particle with spin $\frac{1}{2}$, in Fig. 7.2(a). For such a particle:

$$\mu = g\beta_N\sqrt{3}/2 \quad \text{and} \quad \mu_z = \pm\tfrac{1}{2}g\beta_N \text{ only.}$$

Thus, whichever energy state a spinning nucleus or electron is in, it will always lie more or less across the field and will therefore be under the influence of a couple tending to turn it into the field direction.

Now the behaviour of a spinning nucleus or electron can be considered analogous to that of a gyroscope running in friction-free bearings. Experiments convince us that the application of a couple to a gyroscope does not cause its axis to tilt but merely induces a *precession* of the axis about the direction of the couple.

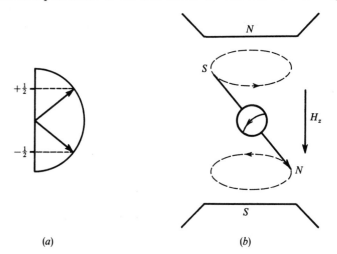

(a)   (b)

Fig. 7.2: (a) The two spin orientations allowed to a nucleus with $I=\frac{1}{2}$, and (b) the Larmor precession of such a nucleus.

Essentially the same occurs with a spinning particle and the precession, known as the Larmor precession, is sketched in Fig. 7.2(b). The precessional frequency (or Larmor frequency) is given by:

$$\omega = \frac{\text{magnetic moment}}{\text{angular momentum}} \times H_z \text{ radians/sec}$$

$$= \frac{\mu H_z}{2\pi I} \text{ c/sec.}$$

Replacing μ and I by their expressions in equations (7.6 and 7.1),

$$\omega = \frac{g\beta_N\sqrt{I(I+1)}}{\sqrt{I(I+1)}\dfrac{h}{2\pi}}\frac{H_z}{2\pi} = \frac{g\beta_N H_z}{h} \text{ c/sec} \qquad (7.12)$$

and, comparing with equation (7.8), we see that the Larmor precessional frequency is just the frequency separation between energy levels.

This, then, is a mechanism by which particle spins can interact with a beam of electromagnetic radiation. If the beam has the same frequency as that of the precessing particle, it can interact coherently with the particle and energy can be exchanged; if of any other frequency, there will be no interaction. The phenomenon, then, is one of *resonance*. For nuclei it is referred to as *nuclear magnetic resonance* (n.m.r.), while for electrons it is called *electron spin resonance* (e.s.r.) or, sometimes, *electron paramagnetic resonance* (e.p.r.).

Experimentally there is a choice of two arrangements. We might either apply a fixed magnetic field to a set of identical nuclei so that their Larmor frequencies are all, say, 60 Mc/sec; if the frequency of the radiation beam is then swept over a range including 60 Mc/sec, resonance absorption will occur at precisely that frequency. On the other hand, we could bathe the nuclei in radiation at a fixed frequency of 60 Mc/sec and sweep the applied field over a range until absorption occurs. The latter arrangement is experimentally simpler and most nuclear (and electron) spin resonance spectra are, in fact, graphs of absorption against applied field.

The probability of transitions occurring from one spin state to another is directly proportional to the population of the state from which the transition takes place. We have seen in the previous section that these populations are very nearly equal, and so during resonance, upward and downward transitions are induced to almost the same extent. However, while the lower state is more populated than the upper (e.g. at equilibrium in the absence of radiation), upward transitions predominate slightly, and a net (but very small) absorption of energy occurs from the radiation beam. When the populations become equal, upward and downward transitions are equally likely, no further absorption can take place and the system is called *saturated*. The equilibrium populations can be re-established if the system loses its absorbed energy; this it cannot do spontaneously, but only as a result of interaction with radiation or with fluctuations in surrounding magnetic fields of the appropriate frequency. The emitted energy can be collected and displayed as an emission spectrum which, since the emission is induced and not spontaneous, is called the *nuclear induction spectrum*.

1.5. Relaxation Times. Let us return to the question posed at the beginning of Section 1.3, and in particular consider the word "immediately". If an external field were suddenly applied to a set of nuclei in bulk material, and if such nuclei can be considered as completely frictionless gyroscopes, then they could not change their orientation in order to produce the correct statistical population of the energy levels unless radiation of the appropriate frequency were present. Without such radiation there would be no mechanism by which the excess energy of the nuclei could be removed from the system (and one would speak of it as having a high "spin temperature"). However, it is a fact that such nuclei do orient themselves to give the appropriate populations of states for a given temperature without the presence of radiation. The mechanism by which excess spin energy is shared either with the surroundings or with other nuclei is referred to generally as a *relaxation process*; the time taken for a fraction $1/e = 0.37$ of the excess energy to be dissipated is called the *relaxation time*.

Two different relaxation processes can occur for nuclei. In the first, the excess spin energy equilibrates with the surroundings (the *lattice*) by spin-lattice relaxation having a *spin-lattice relaxation time* (or *longitudinal relaxation time*) T_1. Such relaxation comes about by lattice motions (e.g. atomic vibrations in a solid lattice, or molecular tumbling in liquids and gases) having approximately the right frequency to interact coherently with nuclear spins. T_1 varies greatly, being some 10^{-2}–10^4 sec for solids, and 10^{-4}–10 sec for liquids, the overall shorter times for liquids being due to the greater freedom of molecular movement leading to larger fluctuations of magnetic field in the vicinity of the nuclei.

Secondly, there is a sharing of excess spin energy directly between nuclei via *spin-spin* (or *transverse*) *relaxation*, the symbol for the time of which is T_2. For solids T_2 is usually very short, of the order 10^{-4} sec, while for liquids $T_2 \approx T_1$.

Values of T_1 and T_2 have a pronounced effect on the width of n.m.r. spectral lines; this is best discussed in terms of the Heisenberg Uncertainty Principle, which states:

$$\Delta E . \Delta t = \frac{h}{2\pi} \approx 10^{-27} \text{ erg. sec}$$

or, dividing through by $h\Delta t$, we have

$$\frac{\Delta E}{h} = \Delta \nu = \frac{1}{2\pi\Delta t} \approx \frac{0.1}{\Delta t} \text{ c/sec,} \qquad (7.13)$$

where ΔE is the uncertainty in energy, or energy spread, of a state, Δv is the corresponding frequency spread, and Δt is the lifetime of the state. When the lifetime of a nuclear spin state is comparatively large, i.e. if *both* T_1 and T_2 are large, then Δv is small, while if *either* T_1 or T_2 is small, the spin state has a short lifetime and Δv is large. Thus for a typical liquid, $T_1 = T_2 = 1$ sec, and hence we calculate $\Delta v = 0.1$ c/sec; while for a solid, $T_2 = 10^{-4}$ (although T_1 is probably larger), and $\Delta v = 1000$ c/sec. Thus n.m.r. experiments are divided into two main classes: broad-line, usually comprising solid samples, and high-resolution, usually of liquids or gases. But it should be noted that while solids rarely, if ever, give high-resolution . spectra, some liquids may give spectra containing broad lines if both T_1 and T_2 are small; such liquids are generally very viscous or contain paramagnetic ions which increase the efficiency of relaxation processes.

2. Nuclear Magnetic Resonance Spectroscopy: Hydrogen Nuclei

We have seen that a particular chemical nucleus placed in a magnetic field gives rise to a resonance absorption of energy from a beam of radiation, the resonance frequency being characteristic of the nucleus and the strength of the applied field. Thus at 15,000 gauss the hydrogen nucleus absorbs at about 64 Mc/sec while in the same field other typical absorption frequencies are, for example: ^{19}F, 60·2; ^{29}Si, 12·8; ^{31}P, 25·8; ^{10}B, 6·9; and ^{11}B, 20·6 Mc/sec. Thus n.m.r. techniques may be used to detect the presence of particular chemical nuclei in a compound and, since, for a given nuclear species, the degree of absorption is proportional to the number of resonating nuclei, to estimate them quantitatively. However, an n.m.r. spectrometer may cost up to some £20,000 and requires a trained operator for its most efficient use; plainly there are many simpler and cheaper methods available to detect the presence or absence of a particular atom in a molecule. Two other characteristics of n.m.r. spectra which have not so far been mentioned make the technique far more powerful and useful; these are the *chemical shift* and the *coupling constant*, which we shall discuss in the following sections.

The vast majority of substances of interest to chemists contain hydrogen atoms and, as this nucleus has one of the strongest resonances, it is not surprising that n.m.r. has found its widest application to these substances. When discussing chemical shifts and nuclear coupling it is convenient to use one type of nucleus as an example, although all spinning nuclei show these phenomena, and in what follows we shall consider the spectra of hydrogen-containing substances only.

   2.1. The Chemical Shift.   Up to now we have considered the behaviour of a single nucleus in an applied field.   Such a situation is not, of course, realizable in practice since all nuclei are associated with electrons in atoms and compounds.   When placed in a

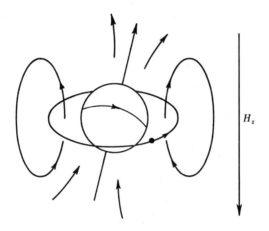

Fig. 7.3: Showing the field, produced by diamagnetic circulation of an electron about a nucleus, which opposes the applied field.

magnetic field the surrounding electrons tend to circulate in such a direction as to produce a field *opposing* that applied (so-called diamagnetic circulation), as shown for a simple atom in Fig. 7.3. Plainly the total field experienced by the nucleus is:

$$H_{\text{effective}} = H_{\text{applied}} - H_{\text{induced}}$$

and, since the induced field is directly proportional to the applied field:

$$H_{\text{induced}} = \sigma H_{\text{applied}}$$

where σ is a constant, we have:

$$H_{\text{effective}} = H_z(1 - \sigma). \tag{7.14}$$

17

Thus the nucleus can be said to be *shielded* from the applied field by diamagnetic electronic circulation; the extent of the shielding will be constant for a given atom in isolation but will vary with the electron density about an atom in a molecule. Thus we may generalize equation (7.14) and write:

$$H_i = H_z(1 - \sigma_i) \tag{7.15}$$

where H_i is the field experienced by a particular nucleus i whose shielding constant is σ_i. As an example, since we know that oxygen is a much better electron acceptor than carbon (since oxygen has the greater electronegativity) then the electron density about the hydrogen atom in C—H bonds should be considerably higher than that in O—H bonds. We would thus expect that $\sigma_{CH} > \sigma_{OH}$ and hence

$$H_{CH} = H_z(1 - \sigma_{CH}) < H_{OH} = H_z(1 - \sigma_{OH}).$$

Thus the field experienced by the hydrogen nucleus in O—H bonds is greater than that at the same nucleus in C—H bonds and, for a given applied field, the CH hydrogen nucleus will precess with a smaller Larmor frequency than that of OH. Put conversely, in order to come to resonance with radiation of a particular frequency (e.g. 60 Mc/sec), a CH hydrogen requires a greater applied field than OH.

It should be noted that electron density is not the only factor determining the value of the shielding constant. Another, frequently very important, contribution to shielding arises from the field-induced circulation of electrons in neighbouring parts of a molecule which gives rise to a small magnetic field acting in opposition to the applied field. This so-called diamagnetic circulation will reduce the net field experienced by a nucleus placed on the axis of the circulating electrons (e.g. as in the group —C≡C—H, where the circulation is around the molecular axis) but will increase the field experienced by a nucleus at right angles to the axis (e.g. in the benzene ring, where the circulation is in the plane of the ring with the hydrogen atoms in this plane). Thus both shielding and deshielding may arise from diamagnetic circulation, and the total of such effects, together with the electron density contribution, is included in the shielding constant σ.

The effect of a steadily increasing field on the energy levels of the CH_3 and OH hydrogen in CH_3OH is shown in Fig. 7.4. The OH

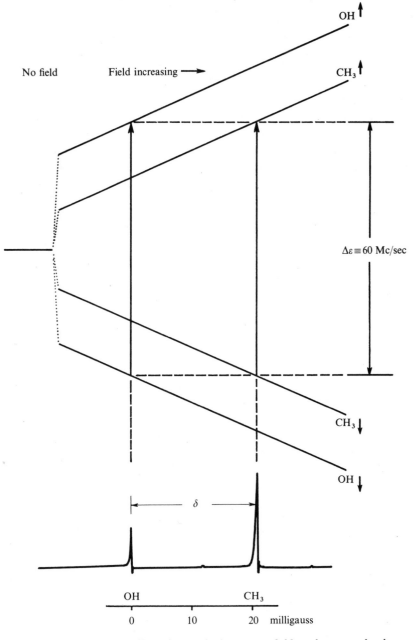

Fig. 7.4: Showing the effect of an applied magnetic field on the energy levels
of the methyl and hydroxyl hydrogen nuclei of methyl alcohol, CH₃OH.
The applied field is increased rapidly initially (dotted portion) until near
resonance at 15,000 gauss, then the increase is much slower. The n.m.r.
spectrum of methyl alcohol is shown at the foot of the diagram.

nucleus, having a smaller shielding constant, experiences a greater field, hence its energy levels are more widely spaced than those of the more shielded CH_3 nuclei at any given applied field. If the system is irradiated with a beam of radiation at, say, 60 Mc/sec while the applied field is increased from zero, the OH nucleus will come into resonance first and absorb energy from the beam, the CH_3 nuclei absorbing at a higher field. This is shown in the spectrum of methyl alcohol, CH_3OH, at the foot of the figure. The fact that the ratio of the absorption intensities (strictly the ratio of the areas under the peaks) is $1:3$ immediately allows us to identify the smaller peak with the single hydrogen nucleus in the OH group, the larger with the CH_3 group. Since neither carbon nor oxygen have nuclei with spin, they do not contribute to the spectrum.

Two very important facets of n.m.r. spectroscopy appear in Fig. 7.4: (i) identical nuclei (i.e. H nuclei) give rise to different absorption *positions* when in different *chemical* surroundings (for this reason the separation between absorption peaks is usually referred to as their *chemical shift*) and (ii) the *area* of an absorption peak is proportional to the number of *equivalent* nuclei (i.e. nuclei with the same *chemical shift position*) giving rise to the absorption. We see here the basis of a qualitative and quantitative analytical technique.

There are several ways in which we can measure the chemical shift, δ, between the absorbance peaks of Fig. 7.4. Firstly, since the spectrum is produced by varying the applied field we can attach a gauss scale (or, rather, a milligauss scale, since the separation is very small) to the spectrum, and we see that $\delta = 20.85$ milligauss. Secondly, remembering that the spectrum could equally have been obtained by varying the frequency at constant field, we might quote the chemical shift in cycles per second using the conversion:

$$15{,}000 \text{ gauss} \equiv 63.87 \text{ Mc/sec}$$

(calculated accurately from equation (7.9)), hence

$$1 \text{ milligauss} \equiv 4.26 \text{ c/sec.}$$

$$\therefore \ \delta = 88.8 \text{ c/sec.}$$

In fact the c/sec scale is used in preference to that in milligauss. We might note at this point that most n.m.r. spectrometers can distinguish lines 0.1 milligauss or 0.4 c/sec apart, which, at an applied field of 15,000 gauss or frequency of 60 Mc/sec, represents a resolving power of some 1 in 10^8; no other spectroscopic technique approaches this precision.

In practice, however, neither of the above chemical shift units is entirely satisfactory although both appear in the older literature. The difficulty can be seen as follows: rewriting equation (7.8) to take account of shielding at nucleus i and combining with equation (7.14):

$$\frac{\Delta E_i}{h} = \frac{g\beta_N H_i}{h} = \frac{g\beta_N H_z(1-\sigma_i)}{h} \text{ c/sec,}$$

and so, in c/sec:

$$\delta = \Delta\varepsilon_{\text{OH}} - \Delta\varepsilon_{\text{CH}_3} = \frac{g\beta_N H_z}{h}(\sigma_{\text{CH}} - \sigma_{\text{OH}}) \text{ c/sec.} \quad (7.16)$$

Now the shielding constant, σ_i, is independent of the applied field or frequency (cf. equation (7.14)) and so plainly chemical shift separation measured in c/sec (or milligauss) is directly proportional to the applied field, H_z: it is, at the least, inconvenient to measure a quantity which changes with the operating conditions of the instrument particularly since different instruments might use fields of anywhere between 6000 and 25,000 gauss.

This difficulty can be overcome if we quote chemical shifts as a *fraction* of the applied field or frequency. Thus for methyl alcohol, $\delta = 88.8$ c/sec at 60·0 Mc/sec is equivalent to $\delta = 1.48 \times 10^{-6}$ or 1·48 p.p.m. (parts per million) of the operating frequency; similarly $\delta = 20.85$ milligauss at 15,000 gauss is also 1·48 p.p.m. of the applied field. If we were now to double the operating field or frequency, the separation *expressed in milligauss or c/sec*, would also double according to equation (7.16), but it would still be just 1·48 p.p.m.

Chemical shift measurements are, of course, formally based on the resonance position of the bare hydrogen nucleus (the proton) as the primary standard; for this there is no shielding and hence $\sigma = 0$. Since this is a quite impracticable standard, it is necessary to choose some reference substance as a secondary standard and to measure the resonance positions of other hydrogen nuclei from this in parts per million; the substance now almost universally selected for hydrogen resonances is tetramethyl silane, $Si(CH_3)_4$, or TMS.

This has several advantages over other substances which have been used as standards:

(*i*) its resonance is sharp and intense since all twelve hydrogen nuclei are equivalent (i.e. have the same chemical environment) and hence absorb at exactly the same position,

(*ii*) its resonance position is to high field of almost all other hydrogen resonances in organic molecules (i.e. σ_{TMS} is large) and hence can be easily recognized,

(*iii*) it is a low boiling point liquid (b.p. 27°C) so can be readily removed from most samples after use.

Thus if a trace of TMS is added to a sample and the complete n.m.r. spectrum produced, the sharp, high-field resonance of the TMS is easily recognized and can be used as a standard from which to calibrate the spectrum and to measure the chemical shift positions of other molecular groupings.   One scale sets the resonance of TMS arbitrarily at zero and counts (in p.p.m.) resonance to *low* field as positive.   This scale, called the δ-scale, has the disadvantage that high numbers imply low-field resonances.   Slightly more convenient, and widely used, is the τ-scale, in which the TMS resonance is set at 10 and low-field resonances are measured in parts per million with decreasing numbers; on this scale some chemical groupings resonate at such low fields that their τ-value is negative, but almost all groups of interest lie between $\tau = 0$ and $\tau = 10$. Figure 7.5(*a*) and (*b*) show both δ- and τ-scales and indicate the approximate resonance positions of some molecules and groups.

In studying this figure we must not lose sight of the fact that the resonances indicated are due to only the *hydrogen* nucleus in the molecule or group concerned.   The nuclei of nitrogen, silicon and phosphorus, which are the only others with spin in the figure, absorb at about 4·6, 12·7 and 26 Mc/sec, respectively, in a 15,000 gauss field—i.e. their "chemical shift" from TMS is some 55, 47 or 34 Mc/sec; one p.p.m. shown in Fig. 7.5 is 60 c/sec.

As one would expect, we note from part (*a*) of the figure that the position of the hydrogen resonance depends upon the atom to which it is directly attached—cf. the series CH_4, NH_3, PH_3, SiH_4 and H_2 between $\tau = 10$ and $\tau = 5·8$.   Far more important, however, is the fact that, when the hydrogen is attached directly to a particular atom, e.g. carbon, its resonance position depends markedly on the nature of the *other* substituents to that atom—cf. the series CH_4,

$CH_3O—$, $C=CH_2$, $O=CH$, etc., in part (b).   When we combine this

with the statement, made earlier, that the area of each resonance is proportional to the number of hydrogen nuclei contributing to that resonance, we see that n.m.r. techniques provide ready means of both qualitative and quantitative group analysis in organic chemistry.   After we have discussed the phenomenon of energy coupling between different nuclei in the following sections, we shall

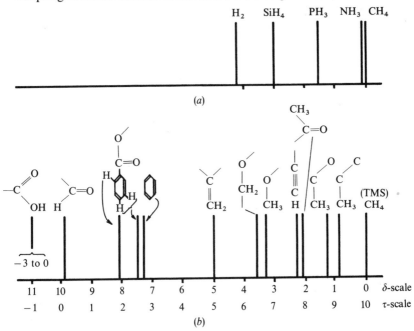

Fig. 7.5: Showing the relationship between the δ- and τ-scales of chemical shift values, and the approximate chemical shifts of some simple molecules and chemical groups.

see that the usefulness of n.m.r. extends even further to the determination of the structure and configuration of molecules.

2.2. The Coupling Constant.   Suppose that two hydrogen nuclei in different parts of a solid, e.g. a crystal lattice, are sufficiently close together in space that they exert an appreciable magnetic effect on each other—in n.m.r. terms "appreciable" means 0·1 milligauss or more.   We show such a case in Fig. 7.6 for two nuclei labelled *A* and X. In (i) of the figure the nuclei are seen to have opposite spin directions while in (ii) they are parallel. Note that we here sketch the z-components of the spin vectors (i.e. the magnetic dipole along the field direction) and that when the applied field H_z is vertically downwards a nuclear spin vector opposed to the field is conventionally said to be "up" and it is of *higher* energy (cf. Fig. 7.1); conversely a vector spin in the field direction is "down" and of lower energy.

In the figure we have drawn the lines of force originating from the magnetic dipole of nucleus A only and we see that when its spin is

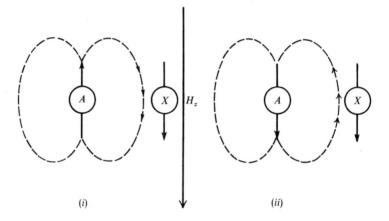

(i)   (ii)

Fig. 7.6: *The direct coupling of nuclear spins. In* (i) *the spin of A increases the net magnetic field at X, while in* (ii) *the field at X is decreased.*

up, as in (i), the field experienced by X is *reinforced* by that from A and hence the energy of X is *lowered*; conversely in (ii) when A is down, the field at X is diminished and its energy raised. If we write ↑↓ and ↓↓ for the spin combination AX in each case, then we have:

$$E_{\downarrow\downarrow}^X > E_{\uparrow\downarrow}^X . \qquad (7.17)$$

Thus we see that nucleus X is made aware of the spin direction of A —the nuclear spins are said to be *coupled*, the strength of the coupling determining the difference between the energy states of X. Normally this coupling is large for hydrogen nuclei (of order 10^4 c/sec when the nuclei are 1 Å apart) and independent of the magnitude of the applied field.   For liquids, as we shall see shortly, molecular tumbling effectively averages such coupling to zero.

If we now imagine the X spin to be reversed (thus increasing its energy by 60 Mc/sec or some other figure depending on H_z), we see that the spin combination ↑↑ (for AX) will raise the energy of X still further while ↓↑ will lower it.   Thus:

$$E^X_{\uparrow\uparrow} > E^X_{\downarrow\uparrow}. \tag{7.18}$$

Comparing equations (7.17 and 7.18) we see that parallel spins raise the energy by coupling, opposed spins lower it.

The influence of A upon X is, of course, precisely reflected by the influence of X upon A and we have immediately:

$$E^A_{\uparrow\uparrow} > E^A_{\uparrow\downarrow} \gg E^A_{\downarrow\downarrow} > E^A_{\downarrow\uparrow}.$$

Thus when two nuclei are directly coupled together a total of eight different energy states arises, four from each nucleus; this is to be compared with the two spin states for each nucleus in the absence of coupling.

For liquids, on the other hand, the direct "across-space" coupling of nuclear spins is not observed because molecular tumbling continually changes the orientations of the molecules and the direct coupling averages exactly to zero.   Now, however, another coupling mechanism becomes important whereby nuclei in the *same molecule* may influence each other.   Consider first Fig. 7.7(*i*) which shows two hydrogen nuclei joined by a pair of bonding electrons, as in the hydrogen molecule.   To a first approximation we can assume that each electron "belongs" to a particular nucleus, so we associate electron (*a*) with nucleus A and electron (*x*) with nucleus X.   Plainly the most stable state energetically is that in which the electronic magnetic dipole is opposed to that of its own nucleus; but theories of chemical bonding tell us that the electrons, which occupy the same orbital, will have their spins opposed also, and so we see that the most stable state will be that in which the nucleus–electron–electron–nucleus spins alternate as shown in the figure; consequently the spins of A and X will preferentially be paired. This is similar to the situation found for the "across-space" effect, but is several orders of magnitude smaller; however, it represents coupling between the nuclei.

Note carefully that the above argument is not intended to be rigorous; we are not saying that the spins of A and X are immutably locked in the paired configuration—in fact the energy difference between paired and opposed configurations is so small that both are virtually equally probable—but the small energy difference is detectable spectroscopically. This coupling mechanism, then,

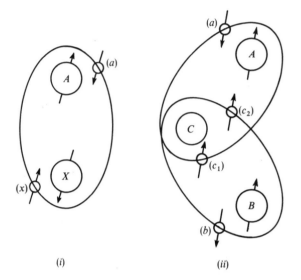

Fig. 7.7: *Showing the coupling of nuclear spins via bonding electrons for* (i) *directly-bonded atoms and* (ii) *atoms bound to a third atom having no spin.*

gives rise to an increase in the stability of the paired configuration with respect to the opposed, i.e.

$$E^A_{\uparrow\uparrow} > E^A_{\uparrow\downarrow} \; ; \qquad E^X_{\uparrow\uparrow} > E^X_{\downarrow\uparrow} \, , \text{ etc.}$$

In Fig. 7.8 we show how the various energy levels of a coupled system are affected by a steady increase in the applied field, and the type of spectrum to be expected. The only selection rule which need be considered is that one nucleus only may change its spin during a transition, and so we have the allowed transitions $E^A_{\downarrow\uparrow} \to E^A_{\uparrow\uparrow}$ (an A transition), or $E^X_{\uparrow\downarrow} \to E^X_{\uparrow\uparrow}$ (an X transition), for instance, but not $E^A_{\downarrow\uparrow} \to E^A_{\uparrow\downarrow}$, etc. There are evidently only two A and two X transitions allowed in this scheme, and each will occur when the separation between the relevant energy levels comes into resonance with the applied radiofrequency beam.

Thus the effect of spin-spin coupling is to split the A and X

resonance peaks into doublets of equal spacing. The separation between the midpoints of the doublets gives the chemical shift between A and X, while the spacing between each pair of lines, measured in c/sec, is called the *coupling constant* and given the

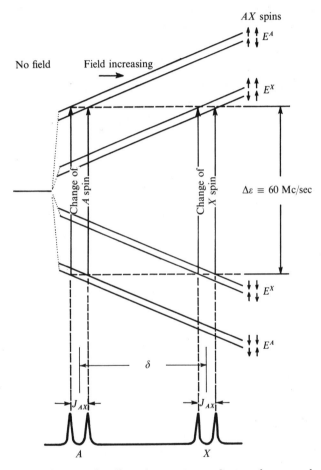

Fig. 7.8: To illustrate the effect of spin-spin coupling on the energy levels of nuclei; as in Fig. 7.4 the applied field is increased rapidly to near resonance.

symbol J or, more specifically, J_{AX} (c/sec are more appropriate units here than p.p.m. since J is field-independent).

In n.m.r. nomenclature the notation AX implies a system containing two nuclei whose chemical shift difference is large compared with the magnitude of the coupling between them. We chose this situation deliberately in the foregoing because, when the chemical shift and coupling constant are comparable in magnitude, the spectrum is not quite so simple as that of Fig. 7.8. A small chemical shift between the two nuclei implies that the energies of their spin "up" states are close together, as are those of their spin "down" states; such a system is called AB, letters close in the alphabet being used for nuclei close in chemical shift. It is a well-established result of quantum mechanics that close energy levels of the same type tend to repel each other; thus it is as if the outer levels in Fig. 7.8 move slightly outwards, and the inner levels move inwards to the same extent. Although the coupling between the nuclei is quite unaffected, the A transitions of Fig. 7.8 will be moved slightly downfield while the X transitions (now relabelled B) will move up-field. Because of this repulsion, then, the chemical shift difference between the nuclei is no longer measured exactly by the separation between the mid-points of the doublets.

A second result of the small chemical shift difference is that the intensities of the absorption lines are perturbed. In the AX system, as shown in Fig. 7.8, all four lines have the same intensity; as the system changes to AB, the centre two lines gain intensity at the expense of the outer two, the total intensity remaining constant. This is illustrated in the typical AB spectrum of Fig. 7.9 which shows part of the n.m.r. spectrum of ethyl cinnamate.

Let us now return to part (ii) of Fig. 7.7. This shows the coupling of hydrogen nuclei which are each attached to a third non-spinning nucleus, such as carbon; an example of this situation is the methylene fragment, $\rangle CH_2$. Now the chain of reasoning runs as follows: the spins of A and (a) are paired, as are those of (a) and (c_1) since both the latter occupy the same orbital; (c_1) and (c_2) have parallel spins, however, since they occupy degenerate orbitals in the same atom (cf. Hund's principle, Section 3.1, Chapter 5); finally (b) and (c_2) are paired and nucleus B has its spin paired with electron (b). We see that the lowest energy state, according to this electron path, is that wherein the spins of A and B are *parallel*, not paired. In this situation the coupling constant, J_{AB}, is defined to be negative, whereas it is defined positive for the previous case shown in Fig. 7.7(i).

Similar arguments to the above indicate that J is again positive for coupling *via* three bonds (e.g. H—C—C—H), and negative *via* four, etc., provided this type of electron path is the predominant contributor to the coupling; in some systems, particularly unsaturated ones (i.e. containing multiple bonds), other electron paths may become important. However, the magnitude of the coupling constant attenuates rapidly with increasing number of bonds; thus for H_2 the coupling (measured indirectly from the spectrum of HD)

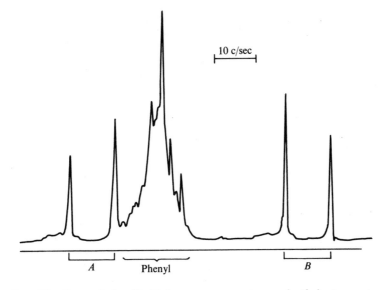

Fig. 7.9: *Part of the 60 Mc/sec. n.m.r. spectrum of ethyl cinnamate,* $C_6H_5CH{=}CH.COOC_2H_5$, *showing the typical AB pattern of the olefinic hydrogens. The broad complex resonance in the centre is due to the phenyl group; the ethyl resonance is off the scale to the right.*

is some 240 c/sec, for the H—C—H fragment (e.g. in methane) it is some 12 c/sec, for H—C—C—H (e.g. in ethanol, CH_3CH_2OH) it is 7 c/sec, and for H—C—C—C—H it is on the present limit of measurement, 0·5 c/sec or less.   However in unsaturated molecules, in which electrons occupy orbitals extending over more than two nuclei and are thus more mobile, the couplings are somewhat larger and have been observed over more than four bonds.

It is not usually easy to determine the sign of a coupling constant experimentally, but where such data is available it is often in good agreement with the alternation theory outlined above.   Further, with this theory calculations can be made predicting the magnitudes of certain couplings; these, too, are in generally good agreement with experimental values.

2.3. Coupling between Several Nuclei.   So far we have considered the effect of coupling between two hydrogen nuclei only; but such nuclei often occur in molecules as groups, particularly CH_3 and CH_2 groups.   We turn now to consider coupling between groups, using the ethyl fragment, CH_3CH_2, as our example.

In the ethyl fragment the three hydrogens of the methyl (CH_3) group have the same chemical shift since all the CH bonds are identical and the shielding at each of the nuclei is the same.   Such nuclei are called *chemically equivalent*.   In the same way the two nuclei of the methylene (—CH_2—) group are chemically equivalent but their chemical shift is, of course, different from that of the methyl nuclei.   Further there is some freedom of rotation of the methyl group in this fragment and hence the interaction between the methyl and methylene nuclei is averaged to the same value—the coupling constant is the same between any methyl and any methylene hydrogen.   The nuclei in the methyl group (or the methylene group) are said to be *magnetically equivalent* as well as chemically equivalent.   This property affords a considerable simplification of the overall spectrum because the *couplings within a group of magnetically and chemically equivalent nuclei do not affect the spectrum* and can be ignored.   In the present example, this means we can neglect coupling between the methyl hydrogens themselves, or between the two methylene hydrogens, and need consider only the coupling between a methyl and a methylene nucleus.   Since all the latter couplings are equal, as explained above, the system has just one J value to be considered.   If the chemical shift between methyl and methylene is large, then we have a system A_3X_2 (where A is a CH_3 hydrogen nucleus, and X a CH_2 nucleus) with a coupling constant J_{AX}.

Consider first the two X nuclei; each has a spin of $\frac{1}{2}$ and we can represent their combined spins as ↑↑, ↑↓, ↓↑ or ↓↓ depending on their separate orientations with respect to the applied field.   A very convenient notation is to write the Greek letter α to represent a spin of $+\frac{1}{2}$ (or ↑) and β for $-\frac{1}{2}$ (or ↓); thus the four spin states of the X_2 group will be $\alpha\alpha$, $\alpha\beta$, $\beta\alpha$ or $\beta\beta$.   Of these, $\alpha\alpha$ represents a total spin of $+\frac{1}{2}+\frac{1}{2}=+1$, $\alpha\beta$ and $\beta\alpha$ each represent total spins of zero, and $\beta\beta$ of -1.   Thus (cf. Fig. 7.6), $\alpha\alpha$ will reinforce the applied field at nuclei A, $\beta\beta$ will reduce it to the same extent, while $\alpha\beta$ and $\beta\alpha$ will

leave it unchanged.   The net result will be a splitting of the resonance of A into three (a triplet) by the three different total spin
orientations of the X nuclei.

The right-hand half of Fig. 7.10(a) illustrates this behaviour and
exemplifies two further points.   Firstly, since the four spin com-
binations of X, $\alpha\alpha$, $\alpha\beta$, $\beta\alpha$ and $\beta\beta$, are all equally likely to occur, the
centre line of the A triplet is twice as intense as the outer lines since
two possible X spin states contribute to it ($\alpha\beta$ and $\beta\alpha$), whereas only
one spin state ($\alpha\alpha$ or $\beta\beta$) contributes to the outer lines.   Secondly,
the diagram shows another useful way in which the effect of coup-
ling can be considered—by imagining the coupling of the A_3 group
to occur to each of the X_2 nuclei *separately*.   Thus one X nucleus
will split the A_3 into a doublet, the second will split each half of this
doublet into a further doublet with the *same spacing* so that the
central lines overlap to give the overall $1:2:1$ triplet.   The separation between these lines is just equal to the coupling constant J_{AX}.

It should now be obvious that the three nuclei of the A group can
take up *four* states with different total spin: $\alpha\alpha\alpha$ (total spin $+\frac{3}{2}$);
$\alpha\alpha\beta$, $\alpha\beta\alpha$, $\beta\alpha\alpha$ (each of total spin $+\frac{1}{2}$); $\alpha\beta\beta$, $\beta\alpha\beta$, $\beta\beta\alpha$ (each $-\frac{1}{2}$); and
$\beta\beta\beta$ ($-\frac{3}{2}$).   Thus the X_2 resonance will be split into a quartet of
lines with intensity ratio $1:3:3:1$.   This is shown to the left of Fig.
7.10(a).   The separation between these lines is again equal to the
coupling constant J_{AX}.

We can now readily generalize the above argument: a group of p
equivalent nuclei will cause the resonance of a coupled nucleus or
group of nuclei to be split into $p+1$ lines; it will itself be split into
$q+1$ lines if the coupled group contains q equivalent nuclei; the
separation between the lines in both multiplets will be equal to J_{pq},
the coupling constant between the groups.   It cannot be too
strongly stressed that it is *not* the number of nuclei in a particular
group which determines the multiplet structure of that group; it is
determined solely by the number of nuclei in the group *to which it is
coupled*.   Thus the triplet pattern of the CH_3 group discussed
above arises because there are *two nuclei in the coupled methylene
group*, not because there are three nuclei in the CH_3 group itself.

The spectrum of ethyl alcohol, CH_3CH_2OH, shown in Fig. 7.10(b)
does not match exactly the theoretical spectrum above it.   In the
first place, lines towards the centre of the ethyl resonance have
gained intensity at the expense of outer lines (cf. the same behaviour
in the AB spectrum of Fig. 7.9) and, secondly, there is evidence of

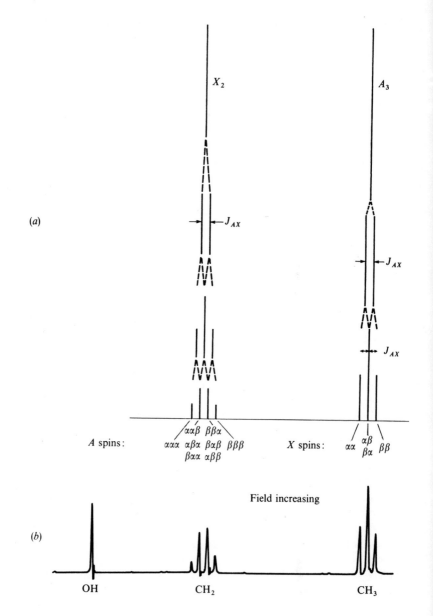

Fig. 7.10: (a) *A theoretical* A_3X_2 *spectrum and* (b) *the actual n.m.r. spectrum of ethyl alcohol,* CH_3CH_2OH *at 60 Mc/sec.*

additional fine structure in the spectrum, particularly in the outer lines of the CH_2 resonance. Both these effects are due to the fact that the theoretical spectrum discussed was A_3X_2 (i.e. very large chemical shift between A and X) whereas the actual system is more properly described as A_3B_2 (chemical shift between CH_3 and CH_2 not much larger than the coupling between them). The observed spectrum can be precisely accounted for when the appropriate values of chemical shift and coupling are taken; the calculations, which involve solving the Schrödinger equation for the system, are rather sophisticated, however, and we shall not discuss them here.

Some other typical coupling patterns are shown in Fig. 7.11. In (a) we see an A_2X_2 spectrum given by the substance cyanhydrin, $CNCH_2CH_2OH$. The methylene groups have different chemical shifts and so it is to be expected that coupling between them will split each resonance into a $1:2:1$ triplet; this is to be observed in the spectrum. Other chemical structures giving rise to similar spectral patterns are 1,2-disubstituted benzenes, or five-membered unsaturated heterocyclic systems such as furan, although in such molecules the spectra are rather more complicated owing to additional couplings which arise between the nuclei. In (b) we show an A_6X spectrum arising from an isopropyl group, $(CH_3)_2CH—$. In this all six methyl hydrogens are equivalent and hence they split the lone methylene hydrogen into a septet, while they themselves are split into a doublet by the single nucleus. Such a pattern is easily recognizable even if the two outer lines of the septet are too weak to be observed, and is very characteristic. Finally in (c) we show the spectrum produced when three different nuclei couple together —an AMX system if all the chemical shifts are large. Within this system there are three different coupling constants: J_{AM}, J_{AX} and J_{MX} so that, for instance, the A resonance is split into a doublet of spacing J_{AM} and then each line of the doublet is further split into a doublet by J_{AX}. Thus each resonance gives rise to a symmetrical quartet as shown in the figure. This pattern might arise from the three ring hydrogens of a monosubstituted furan or similar molecule (in this example we show the spectrum of an α-furan) or it might be from a vinyl group $CH_2{=}CH—$, where again all three hydrogen nuclei have a different chemical shift.

These coupling patterns, and any others derived by the methods outlined previously for the $CH_3CH_2—$ fragment, are considerably complicated if the chemical shift between coupled nuclei is not

18

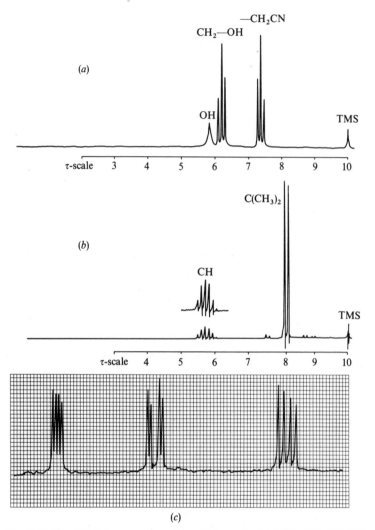

Fig. 7.11: The 60 Mc/sec n.m.r. spectra of (a) cyanhydrin, $CN.CH_2CH_2OH$, (b) isopropyl iodide, $(CH_3)_2CHI$, and (c) 2-furanoic acid,

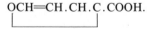

$$OCH=CH.CH.C.COOH.$$

In (a) and (b) the normal τ-scale, in p.p.m. from tetramethyl silane (TMS), is shown, while in (c) the scale is expanded to 1 c/sec per division in order to show clearly the three quartets. The resonance of the COOH group is thus off the scale to the left.

large. Usually, however, the additional fine structure produced and the intensity perturbations do not prevent recognition of the overall pattern, particularly when some experience has been gained from studying actual spectra. The tremendous analytical value of such patterns is obvious since their recognition immediately gives information about the chemical groupings present in the molecule under examination. We discuss this aspect of n.m.r. spectra more fully in the next section.

2.4. Chemical Analysis by N.M.R. Techniques. In the preceding sections we have built up a picture of the application of n.m.r. to constitutional and structural studies. Thus the observation of the τ-values of lines in a spectrum (or of the centres of multiplets if coupling is occurring) immediately indicates, with very little ambiguity, the types of hydrogen-containing groups within the molecule, while the relative intensities of the lines yield directly the proportions in which these groups occur.

Further, the multiplet structure of each group in the spectrum gives information on the number of hydrogen nuclei coupled to that group and in this way shows which groups are near neighbours in the molecule. Thus groups such as CH_3CH_2, $—CH_2CH_2—$, $(CH_3)_2CH—$, etc., can be instantly recognized from the n.m.r. spectrum.

As an example of the use and limitations of n.m.r. spectroscopy in analysis, consider the spectrum shown in Fig. 7.12. Resonances are centred at $\tau = 1.8$, 2.5, 5.6 and 8.7, the position of the former two and the coupling pattern of the latter two suggesting that they

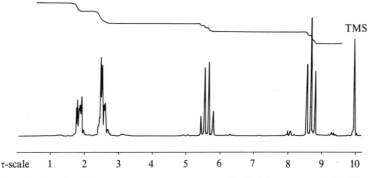

τ-scale   1   2   3   4   5   6   7   8   9   10

Fig. 7.12: The 60 Mc/sec n.m.r. spectrum of ethyl benzoate, $C_6H_5CO.$
OCH_2CH_3, to illustrate the use of n.m.r. as an analytical technique.

arise from a phenyl group (C_6H_5—) and an ethyl group (C_2H_5—) respectively. The integral trace on the spectrum shows these resonances to have relative intensities of 5:2:3 (the two phenyl resonances here being summed) so we know the phenyl and ethyl to be present in 1:1 ratio. If we also know that the molecular formula of the substance is $C_9H_{10}O_2$, we can rapidly deduce that the resonances at $\tau = 1.8$ and 2.5 are due to hydrogen nuclei respectively ortho and meta/para to a carbonyl group (cf. Fig. 7.5), while those at 5.6 and 8.7 are consistent with the grouping $CH_3CH_2.O.CO$—. The molecule is thus ethyl benzoate, $C_6H_5CO.O.CH_2CH_3$.

Note that the n.m.r. spectrum, while allowing us to deduce directly the presence of phenyl and ethyl groups, does not indicate the *presence* of groups not containing magnetic nuclei, in this case O and CO. However, once these groups are known to be present, either by determination of the molecular formula or by observation of the infra-red spectrum, then the n.m.r. spectrum does indicate their *position* in the molecule. Thus the coupling pattern shows clearly that the CH_3 and CH_2 groups are directly bound and not, for example, joined via O or CO: $CH_3.O.CH_2$— or $CH_3.CO.CH_2$—, since in these latter configurations the coupling constant would be immeasurably small.

This very simple example serves to indicate the method of approach when using n.m.r. for analytical purposes. Of course in more complex cases complete structural determinations are seldom possible from the n.m.r. spectrum alone, but when taken in conjunction with other techniques, in particular infra-red spectroscopy, a great deal of useful structural information can usually be obtained about an unknown molecule.

2.5. Exchange Phenomena: Double Resonance. The student may have been puzzled by one aspect of the alcohol spectra shown in Figs. 7.4, 7.10 and 7.11; in these three spectra the resonance of the —OH hydrogen is shown as a single line whereas we might now expect it to be coupled with neighbouring CH_3, CH_2 or CH nuclei and hence have multiplet structure—quartet, triplet or doublet, respectively. The reason it does not so couple is attributable to the fact that the hydroxyl hydrogen is readily exchanged with other hydrogen nuclei or ions in its vicinity; when this happens, the replacement hydrogen does not necessarily have the same spin direction as that being displaced and, if the exchange occurs sufficiently rapidly, the neighbouring CH nucleus experiences a "coupling field" which is averaged to zero. Thus there is no net coupling between the OH and neighbouring groups.

If the exchange is prevented by rigorous drying of the alcohol samples, coherent coupling appears and the OH resonance has the expected multiplet structure. It can be shown theoretically that for a coupling constant of J c/sec the coupled nucleus must exchange more rapidly than $J/2\pi$ times per second for the multiplet to collapse to a singlet. For the alcohols considered above $J \approx 6$ c/sec, and hence an exchange rate of only about once per second is sufficient to destroy coherent coupling. Of course, the transition from coupling to no coupling is not abrupt; at exchange rates rather lower than $J/2\pi$ the lines of the OH multiplet begin to broaden, at $J/2\pi$ they are so broad that all trace of line splitting is obliterated and, at higher rates of exchange, the broad line sharpens until the single very sharp resonance of the illustrated spectra are seen.

An obvious application of this effect is to the study of hydrogen exchange kinetics; exchange rates frequently vary with temperature or concentration and such variations can be followed very precisely by observation of the change in line shape of n.m.r. signals.

A way in which coherent internuclear coupling can be destroyed without actual physical exchange of the coupled nucleus, is by sufficiently rapid reversal of its *spin*. As we have seen, the production of an n.m.r. spectrum depends on the interaction between a nuclear spin and incident radiation. When the radiation is of the appropriate (Larmor) frequency the interaction causes spin reversal leading to net absorption of energy from the beam. If we have a substance containing two coupled nuclei having different resonance frequencies, v_1 and v_2, in a given applied field, we can observe the spectrum of, say, nucleus 1, by applying a radiation beam of frequency v_1. If *at the same time* we also apply strong radiation at frequency v_2 we shall cause rapid transitions between the spin states of nucleus 2. The detector, being tuned to frequency v_1, will not show a spectrum for nucleus 2, but the "stirring" of nuclear spin which results destroys coherent coupling between nuclei 1 and 2; consequently the multiplet resonance of 1 collapses to a singlet.

This technique, which is known as *double irradiation* or *double resonance*, is very useful for simplifying complex spectra. When a spectrum arises from several different nuclear groups coupled together in a complicated way, the resulting pattern is likely to be very difficult to understand. If each group in turn is subjected to double resonance, however, while the rest of the spectrum is scanned in the usual way, any coupling between the doubly irradiated group and the other groups is destroyed and their multiplet structure is considerably simplified. Not only may the simpler spectra be directly amenable to analysis, but by noting which multiplets collapse, we know immediately which groups are directly coupled to that irradiated.

Nuclei in different chemical surroundings (i.e. having different chemical shifts) may, in addition, have their chemical shift positions averaged by exchange phenomena. If the exchange is considerably more rapid than the difference between the two chemical shifts (expressed in c/sec) then only one sharp resonance signal will appear, midway between the separate chemical shift positions; for a much slower exchange, two sharp resonances will be observed at the proper positions, while for intermediate rates either one or two broad resonances occur.

The exchange giving rise to the averaging may be a physical exchange of the nuclei (as in the exchange of the OH proton in alcohols considered earlier) or merely an internal rearrangement, such as the rotation of a methyl group or the interconversion of two chair forms of cyclohexane. In the latter, interconversion results in all the equatorial hydrogens becoming axial, and vice versa; experiments on substituted cyclohexanes, in which the interconversion is inhibited, show that equatorial and axial protons have chemical shifts differing by up to some 40 c/sec, but in the unsubstituted compound only one sharp resonance line is observed, showing that the interconversion takes place considerably more rapidly than 40 times per second.

3. Nuclear Magnetic Resonance Spectroscopy: Nuclei other than Hydrogen

3.1. Nuclei with Spin $\frac{1}{2}$. In general, any nucleus with a spin of $\frac{1}{2}$ will give rise to n.m.r. signals provided the appropriate magnetic field and radiofrequency are applied. There are many such nuclei, of which probably ^{13}C is the most important for the general chemist; this nucleus has not been widely studied, however, since it gives rise to extremely weak signals and, even after considerable concentration of the isotope above its naturally occurring 1% level, satisfactory n.m.r. spectra are not easy to obtain. Nuclei more often studied include ^{19}F and ^{31}P, both being the only isotopes of their particular elements, ^{29}Si (4·7%) and various metals such as ^{117}Sn, ^{119}Sn, ^{195}Pt, ^{205}Tl or ^{207}Pb.

For all these nuclei the phenomena of chemical shift and spin-spin coupling are observed in the spectra and generally both are considerably larger than their hydrogen counterparts. Thus, while a range of some 15 p.p.m. contains virtually all the known hydrogen chemical shifts, values for phosphorus-containing compounds span some 400 p.p.m., fluorine some 600 p.p.m. and a few metals, notably ^{205}Tl and ^{207}Pb range over 14,000 p.p.m. or more. This behaviour can be traced to the greater number and increased mobility of the extra-nuclear electrons leading to greater variation in diamagnetic shielding.

The increase in spin-spin coupling constants is probably attributable to the same cause although the increase is not so marked as in the case of chemical shifts. Thus the coupling between two directly bonded phosphorus nuclei is some 600 c/sec, while it may be as high as 1400 c/sec for phosphorus bonded to fluorine. These figures, however, are only some 3–6 times larger than the corresponding direct H—H coupling of 240 c/sec. A further general point is that couplings tend to attenuate less rapidly with increase in the number of bonds separating the coupled nuclei than do those of hydrogen.

Two practical advantages follow from the large chemical shifts of these nuclei. Firstly, less precise instrumentation is required in order to obtain data useful for structural determinations since the tolerances with which chemical shifts must be measured are correspondingly larger. Secondly, the spectra obtained are simpler to analyse in that they have less of the complications which arise in hydrogen spectra when the chemical shifts are comparable in magnitude with coupling constants. The spectra are thus of the $A_aM_mX_x$... type, rather than $A_aB_bC_c$....

We shall not discuss the spectra of these nuclei further here except to state that a great deal of useful structural information is obtainable from them. For a more detailed account the reader is referred to Chapter 12 of the book by Pople, Schneider and Bernstein, mentioned in the bibliography at the end of this Chapter.

3.2. Nuclei with Spin Greater than $\frac{1}{2}$. We saw at the beginning of this chapter that the application of a magnetic field to any nucleus with spin I causes the spin vector to become oriented in any one of $2I + 1$ possible directions, each associated with a slightly different energy level. Thus Fig. 7.1 shows the situation for a nucleus with a spin of 1, e.g. ^{14}N. Since, in a given field, the spacing between the energy levels of a particular nucleus are all identical, transitions induced between any neighbouring levels will result in the emission or absorption of energy at the same frequency. Thus only one resonance line will appear for each nucleus. Since the only transitions allowed are for $\Delta I_z = \pm 1$, "overtone" transitions do not occur in n.m.r.

In principle, then, *any* spinning nucleus will give rise to a single resonance line when the appropriate field and frequency are applied; the position of the resonance for a given nucleus will vary with its chemical surroundings and it will be split into a multiplet by inter-action with other spinning nuclei—in other words, the phenomena of chemical shift and spin-spin coupling will be observed. In practice, however, the spectra of nuclei with spin greater than $\frac{1}{2}$ are usually intrinsically weak and not easy to observe and they are not much used, in themselves, for analytical or structural studies. However, elements such as ^{14}N (spin $I = 1$), and the halogens, chlorine, bromine (both of spin $= \frac{3}{2}$) and iodine (spin $= \frac{5}{2}$) occur widely in chemistry, and we should consider what effect the presence of these nuclei might have, by virtue of spin-spin coupling, on the n.m.r. spectrum of neighbouring *hydrogen* nuclei.

Let us take the ^{14}N—H group as an example; we would expect the coupling here to be reasonably large because the two nuclei are directly bonded (J, in fact, is observed to be some 50 c/sec for this group). Now the unit spin of the nitrogen nucleus can take up one of three orientations in an applied field (cf. Fig. 7.1), which we may represent as ↑, → and ↓. Comparing with Fig. 7.6 we see that the spin direction ↑ will reinforce the field at the hydrogen nucleus, ↓ will reduce it to the same extent, while it is plain that → will leave it unaltered. We expect the hydrogen resonance to be split into a *triplet*, therefore, and, since all three spin orientations of the nitrogen are equally likely, each line of the triplet will have the same intensity. The formation of this $1:1:1$ triplet is shown schematically in Fig. 7.13(*a*). In Fig. 7.13(*b*) we reproduce the spectrum of acidified methylamine, CH_3NH_2, to show the triplet structure of the hydrogen resonance. The solution is acidified merely to suppress the otherwise rapid exchange of the hydrogens attached to nitrogen,

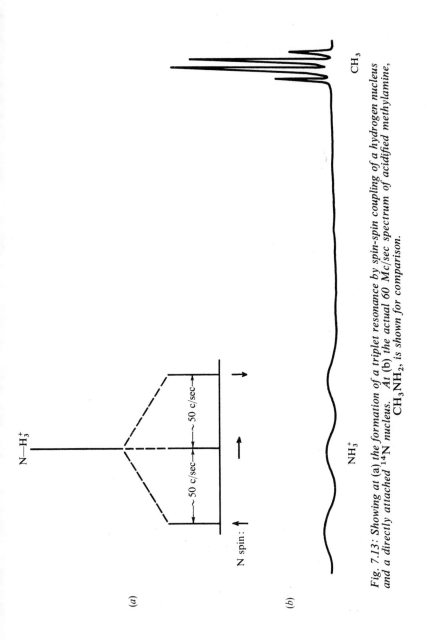

Fig. 7.13: Showing at (a) the formation of a triplet resonance by spin-spin coupling of a hydrogen nucleus and a directly attached ^{14}N nucleus. At (b) the actual 60 Mc/sec spectrum of acidified methylamine, CH_3NH_2, is shown for comparison.

which exchange would destroy the N—H coupling. One result of this is that the molecule becomes converted to the methylammonium *ion*, $CH_3NH_3^+$, and so the methyl resonance is split into a 1:3:3:1 quartet by spin-coupling with the $—NH_3^+$ hydrogen nuclei; the fact that this quartet is sharply defined implies that the exchange rate is small.   In spite of this we note that the NH triplet consists of very broad resonance lines; this broadening is due to quadrupole relaxation which we shall discuss shortly.

We can easily generalize the above discussion to other nuclei.   A nucleus with spin I can take up one of $2I+1$ equally likely spin orientations in an applied field; of these, some will reinforce and some reduce the field experienced by a neighbouring nucleus, so the resonance of the neighbour will be split into a multiplet of $2I+1$ lines with equal intensity.   Thus a single chlorine nucleus would tend to produce quartet structure in its neighbours (spin $=\frac{3}{2}$), while a single iodine nucleus ($I=\frac{5}{2}$) would produce sextets.   In practice, however, these splittings are not observed because of quadrupole relaxation which causes rapid transitions between the spin states (see next section).

3.3. Quadrupole Effects.   In addition to the magnetic moment discussed throughout this chapter, all nuclei with a spin $\geqslant 1$ also possess an *electric quadrupole moment*, which arises because the nuclei are not spherical.   Such nuclei, in fact, are shaped either like a symmetrical egg or like a tangerine (elongated or flattened at the poles, respectively).   Even if the charge density within the nucleus is constant, the distorted shape gives rise to a charge distribution which is non-spherical; the electric quadrupole moment is a measure of the departure from sphericity, being positive for egg-shaped and negative for tangerine-shaped nuclei.   For spherical nuclei (i.e. spin $\frac{1}{2}$ or 0) the electric quadrupole moment is zero.

The electric moment interacts strongly with an applied electric field and, if such a field has a pronounced gradient at the nucleus, the nuclear moment will tend to lie in the field direction; further if the gradient changes direction, the nuclear moment will try to follow this change.   Thus, consider the pyramidal molecule of ammonia, NH_3.   This has the three hydrogen nuclei arranged at the corners of the base of a pyramid, the nitrogen nucleus being above and equidistant from them.   The presence of three positively charged nuclei on one side of the nitrogen nucleus produces a strong electric field gradient at that nucleus which tends to orient its quadrupole moment axis perpendicularly to the plane containing the hydrogen nuclei. If, due to molecular tumbling, the molecule rotates as a whole, the nitrogen nucleus tends to follow this rotation. So strong is the coupling between nuclear electric moment and field gradient that it is sufficient to prevent the coherent alignment of the nuclear *magnetic* moment in an applied magnetic field—in other words the quadrupole electric moment supplies a mechanism by which the spin orientation may be relaxed. Here, then, we have another—and usually very efficient—relaxation mechanism (cf. Section 1.5) called *quadrupole relaxation*.

If, on the other hand, we consider the ammonium ion, NH_4^+, which consists of a nitrogen nucleus at the centre of a regular tetrahedron of hydrogen nuclei, there is no field gradient at the nitrogen nucleus. Thus its spin can be oriented by an applied magnetic field quite independently of the orientation of the hydrogen nuclei—the quadrupole relaxation is extremely weak in this case. We see, then, that the efficiency of quadrupole relaxation will depend upon the symmetry of the surroundings. Univalent atoms, such as the halogens, will always be situated in a field gradient and will experience efficient relaxation; the relaxation of polyvalent atoms, such as nitrogen, will vary from molecule to molecule.

Quadrupole relaxation has two effects in n.m.r. spectroscopy. The first of these, the broadening of the n.m.r. signal from the nucleus possessing a quadrupole moment, is very similar to the broadening caused by fast relaxation (cf. Section 1.5). When quadrupole relaxation occurs the lifetime of a particular spin state is very short, of the order of 10^{-4} sec or less. If we put $\Delta t = 10^{-4}$ in equation (7.13), we have a frequency spread of about 1000 c/sec; thus the resonance line of a nucleus with a short relaxation time may be extremely broad and difficult to detect. As a comparison, a similar calculation for hydrogen nuclei (where, of course, quadrupole relaxation cannot occur) would put $\Delta t \approx 1$ sec, and hence $\Delta v \approx 0.1$ c/sec. This, being rather less than the resolving power of n.m.r. spectrometers, means that hydrogen resonances are usually extremely sharp and narrow.

For the second effect of quadrupole relaxation we recall the discussion of double resonance in Section 2.5, wherein an experimental method for "stirring" or relaxing coupled spins was described. In quadrupole relaxation we have another method for destroying coherent coupling between nuclei. If the relaxation is highly efficient, as for example in halogen nuclei, the "stirring" is so rapid that coupling is completely destroyed and the resonances of neighbouring nuclei remain sharp; for less efficient relaxation, as in many nitrogen-containing molecules, the coupling is only partially destroyed and multiplet but broad resonances result from neighbouring nuclei. An example of this has already been seen in the very broad triplet formed by the $—NH_3^+$ group of the methylammonium ion of Fig. 7.12. In routine n.m.r. spectroscopy, the only commonly occurring quadrupolar nucleus whose effect is noticeable in spectra is, in fact, that of ^{14}N: here the effect is usually to broaden the resonance of neighbouring nuclei, sometimes with the appearance of multiplet structure. Often the broadening is so great that the hydrogen resonances disappear completely into the background noise.

A single crystal of a solid substance has a well-ordered and regular structure, apart from possible lattice defects, and any quadrupolar nucleus contained in a crystal will find itself in exactly identical surroundings as similar nuclei in other regions of the lattice. Thus all the quadrupole moments will tend to lie in the same direction, insofar as this is consistent with the Boltzmann energy distribution. Even without the application of an external magnetic field such nuclei will be able to absorb energy coherently from a beam of radiation at an appropriate frequency. These absorptions, which occur in the region 1–1000 Mc/sec, are known as the *nuclear quadrupole resonance* (n.q.r.) spectrum or sometimes the *pure quadrupole spectrum* to emphasize the fact that an external field is not applied. N.q.r. spectroscopy essentially uses the quadrupolar nucleus as a probe to detect and estimate electric field gradients in the crystal and the data obtained is invaluable in applications of crystal field theory. The spectra are not easy to interpret, however, and will not be discussed here; the interested reader is referred to Chapter II of the book by Nachod and Phillips listed in the bibliography.

4. Electron Spin Resonance Spectroscopy

4.1. Introduction. In Chapter 6 we saw that the majority of stable molecules are held together by bonds in which electron spins are opposed; in this situation there is no net electron spin, no electronic magnetic moment and hence no interaction between the electron spins and an applied magnetic field. On the other hand some atoms and molecules contain one or more electrons with unpaired spins and these are the substances which are expected to show *electron spin resonance* (e.s.r.) spectra; since such substances show bulk paramagnetism, this type of spectroscopy is often referred to as *electron paramagnetic resonance* (e.p.r.).

Substances with unpaired electrons may either arise naturally or be produced artificially. In the first class come the three simple molecules NO, O_2 and NO_2, and the ions of transition metals and their complexes, e.g. Fe^{3+}, $[Fe(CN)_6]^{3-}$, etc. These substances are stable and easily studied by e.s.r. Unstable paramagnetic materials, usually called *free radicals* or *radical ions*, may be formed either as intermediates in a chemical reaction or by irradiation of a "normal" molecule with ultra-violet or X-ray radiation or with a beam of nuclear particles. Provided the lifetimes of such radicals are greater than about 10^{-6} sec they may be studied by e.s.r. methods; shorter-lived species may also be studied if they are produced at low temperature in the solid state—so-called *matrix techniques*—since this increases their lifetimes.

As in all forms of spectroscopy, four properties of the spectral lines are of importance, viz. their intensity, width, position, and multiplet structure. The first two we can deal with quite briefly, the final pair merit separate sections.

The *intensity* of an e.s.r. absorption is proportional to the concentration of the free radical or paramagnetic material present. Thus we have immediately a technique for estimating the amount of free radical present; the method is extraordinarily sensitive, in favourable cases some 10^{-13} mole of free radical being detectable.

The *width* of an e.s.r. resonance depends, analogously to n.m.r., on the relaxation time of the electronic spin state. Electron spin relaxations are much more efficient than those of nuclei and relaxation times of the order of 10^{-7} sec or less are usual. Applying the Heisenberg uncertainty relation, equation (7.13), we find that line widths are of the order 10^6–10^7 c/sec, or 1–10 Mc/sec; when expressed in units of the applied magnetic field this is of the order of one gauss. In e.s.r. work such lines are called "narrow" although they are, of course, extremely wide when compared with n.m.r. line widths (some 0·02 milligauss).

4.2. The Position of E.S.R. Absorptions; the g Factor. We know from Section 1.2 that the spin energy levels of an electron are separated in an applied magnetic field, H_z, by an amount:

$$\frac{\Delta E}{h} = \frac{g\beta H_z}{h} \text{ c/sec,} \qquad (7.19)$$

where β is the Bohr magneton $(9\cdot273 \times 10^{-21}$ erg. gauss$^{-1})$ and g the Landé splitting factor.   A resonance absorption will thus occur at a frequency $v = \Delta E/h$ c/sec.   From equation (7.19) we see that the position of absorption varies directly with the applied field and, since different e.s.r. spectrometers operate at different fields, it is far more convenient to refer to the absorption in terms of its observed g value.   Thus, rearranging equation (7.19) we have:

$$g = \frac{\Delta E}{\beta H_z} = \frac{hv}{\beta H_z} \qquad (7.20)$$

and if, for example, resonance were observed at 8388·225 Mc/sec in a field of 3000 gauss, it would be reported as resonance at a g value of 2·0023.   This very precise figure is the g factor for a free electron (rather than the slightly approximate value of two given by putting $L=0$ in equation (5.28)), but it is a remarkable fact that virtually all free radicals and some ionic crystals have a g factor which varies only some $\pm 0\cdot003$ from this value.   The reason for this is essentially that in free radicals the electron can move about more or less freely over an orbital encompassing the whole molecule (as we shall see in the next section) and it is not confined to a localized orbital between just two of the atoms in the molecule.   In this sense it behaves in very much the same way as an electron in free space, having $L=0$.

Some ionic crystals, on the other hand, have very different g factors, values between about 0·2 and 8·0 having been reported. The difference here is that the unpaired electron is contributed by, and "belongs" to, a particular atom in the lattice, usually a transition metal ion.   Thus the electron is localized in a particular orbital about the atom, and the orbital angular momentum (L value) couples coherently with the spin angular momentum giving rise to a g value consistent with equation (5.28).

Nonetheless, many ionic crystals show a g factor very close to the free electron value of 2; this may come about in two ways:

(i) the ion contributing the electron may exist in an S state (i.e. $L=0$).   For example, the ground state of Fe^{3+}, in which five d electrons are unpaired (i.e. $S=\frac{5}{2}$, $2S+1=6$), has zero orbital momentum. Thus $L=0$, $J=S+L=S$, and the term symbol is $^6S_{5/2}$ (cf. Chapter 5). Since $J=S$, $g=2$ (equation (5.28)).

(*ii*) The electric fields set up by all the ions in a crystal may be sufficiently strong to *uncouple* the electron's orbital momentum from its spin momentum—i.e. coherent Russell–Saunders coupling breaks down and, on the application of a magnetic field, the electron spin vector precesses *independently* about the field direction. Thus the value of L is immaterial and the g factor reverts to two. On the other hand, if the internal crystal field is weak, or if the paramagnetic electron is well-shielded from the field (e.g. as in a rare earth metal, where the relevant electron orbit is buried deep within outer electron shells), L and S couple to produce a resultant J which itself precesses about the applied magnetic field, and g is given by equation (5.28). Intermediate cases also occur where L and S are only partly uncoupled, the residual orbital contribution to the energy giving rise to a g value not easily predictable theoretically.

4.3. The Multiplet Structure of E.S.R. Absorptions. In e.s.r. spectroscopy we must distinguish between two kinds of multiplet structure; there is the *fine structure* which occurs only in ionic crystals, and the smaller *hyperfine structure* occurring both in crystals and in free radicals.

Fine structure arises when the paramagnetic material has more than one unpaired electron, since then the total spin S can have more than one value. Thus for two electrons we could have either $S=1$ or $S=0$. The different spin states interact differently both with the internal electric crystal field, and with any externally applied magnetic field. For our example of two unpaired spins we have the situation shown in Fig. 7.14; on the left we see the splitting between the levels $S=1$ and $S=0$ by the crystal field (the so-called zero-field splitting, since $H_z=0$), and on the right the effect of gradually increasing the magnetic field: the $S=0$ state is unchanged, while the $S=1$ splits into $S_z=+1$ and $S_z=-1$ levels. If the system is irradiated with microwave radiation of frequency v, transitions will occur under the selection rule $\Delta S_z=\pm 1$ and we see that in this case two such transitions are possible. The fine structure spectrum will thus consist of two absorptions whose separation will be of the order of hundreds of gauss. In general n unpaired spins will result in a fine structure spectrum of n lines.

Hyperfine structure, on the other hand, arises through coupling of the unpaired electron with neighbouring nuclear spins in much the same way as the coupling between nuclear spins discussed in Section 2.2.   In general a nucleus with spin I will split the resonance line of an electron into a multiplet with $2I+1$ lines of equal intensity.   The separation between the lines is usually of the order of 1–20 gauss (i.e. some 50 Mc/sec) which is larger by a factor of approximately 10^6 than nucleus–nucleus coupling.   The reason is

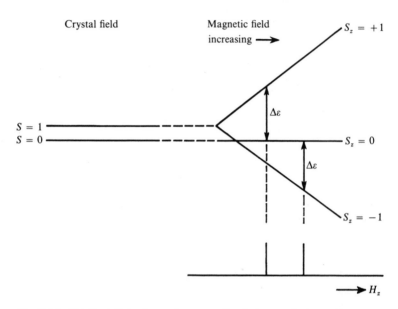

Fig. 7.14: *On the left is shown the separation between electron spin energy levels caused by the internal crystal field, and on the right the additional separation caused by the application of a steadily increasing magnetic field.*

that an electron can approach a nucleus much more closely than can another nucleus, and so will interact more strongly with it.

The biggest factor influencing the magnitude of electron–nucleus coupling is the amount of time which the electron spends in the vicinity of the coupled nucleus or, in other words, the *electron density* at the nucleus.   We can express this as:

$$A = R\rho \qquad (7.21)$$

where A is the observed coupling, ρ the electron density, and R the intrinsic coupling for unit density.   Thus for the hydrogen atom in the ground state the observed hyperfine splitting is some 500 gauss (incidentally the largest known) and, since the electron density in the $1s$ orbital must be unity (the electron is nowhere else), R is also 500 gauss. Now in the methyl radical, $\cdot CH_3$, the electron resonance is observed to have a quartet structure with lines of intensity ratio $1:3:3:1$ and a separation of 23 gauss. The quartet pattern is consistent with the interaction of the electron *equally* with all three hydrogens (cf. the similar pattern for the CH_2 group coupled to CH_3 in the n.m.r. spectrum of Fig. 7.10), and we calculate from equation (7.21) that $\rho = 23/500 = 0.046$. This implies that the electron spends some 5% of its time in the $1s$ orbital of each hydrogen, the remaining 85% in the neighbourhood of the carbon atom.

We can take this argument further. The coupling between each ring hydrogen and the unpaired electron in the free radical p-benzosemiquinone:

is about 2·4 gauss. This tells us, firstly, that the electron density at each hydrogen is $2\cdot4/500 \approx 0.005$. Now in the methyl radical, considered above, the coupling between electron and hydrogen when the electron density at the carbon is 0·85, is 23 gauss; it follows that the electron density at each of the ring carbon atoms in p-benzosemiquinone must be just one-tenth of this, 0·085, in order to give a coupling one-tenth as large. Thus we know the electron density at each atom of the molecule: each H, 0·005; each C, 0·085; each O (by difference), 0·24, or $\frac{1}{2}$%, 8·5% and 24% of the electron's time is spent at each nucleus, respectively.

In this way e.s.r. techniques help us to build up a picture of the electron distribution within a molecule; the results obtained are in good agreement with the predictions of molecular orbital theory. They also help in the understanding of some chemical reactions—a positively charged reactant will plainly tend to attack that part of a molecule where the electron density is greatest, and vice versa.

5. Techniques and Instrumentation

Basically n.m.r. and e.s.r. spectrometers require components with functions similar to those already described for other spectroscopic techniques, but with the important addition of a powerful magnet. For n.m.r., particularly, we have seen that a resolving power of 1 c/sec requires that the magnetic field be controlled to within one part in 10^8 or better (i.e. better than 0·1 milligauss in 10,000 gauss); in addition the field must be the same over the whole sample under test—that is, an area of some 2×0.5 cm. These requirements are very stringent. Good homogeneity is achieved by using magnets with pole pieces some 20 cm in diameter and placing the sample near the centre, by having highly polished pole faces, by using special electric "shims" round the magnet through which small currents are passed to offset any departure from homogeneity, and finally by spinning the whole sample about an axis at right angles to the field direction thus effectively averaging the field experienced by each part of the sample.

For n.m.r. work the magnet may be either of the electromagnetic or of the permanent type; both have advantages and disadvantages. An electromagnet is the more flexible and changes in the field strength can be made in a matter of minutes for the study of different nuclei. However it requires very complex electronics to maintain a current of sufficient stability through the coils to meet the 1 in 10^8 requirement. Further the heat dissipated in the magnet windings is very large, water cooling is essential, and it is not easy to thermostat the magnet accurately. While the temperature of a permanent magnet is usually controlled to very fine limits, different nuclei can be studied only by a change in the operating *frequency* of the instrument and thus a separate crystal oscillator is required for each nucleus of interest—the normal laboratory will have three oscillators, one each for hydrogen, fluorine and phosphorus.

Whether the magnet is permanent or electric the spectrum is usually scanned by passing a small current through "sweep" coils arranged round the pole pieces, thus producing a steady change in the field.

We cannot leave the subject of n.m.r. magnets without a mention of the latest advance in technique—the introduction of the super-conducting magnet. When metals are cooled to temperatures less than a degree above absolute zero their electrical resistance falls to zero and they are said to be superconducting. A current once started in a coil of such material will continue to circulate until the metal is allowed to warm up. Normal electromagnets have a maximum operating field of about 25,000 gauss, but there seems to be no limit, at present, to the field possible with a magnet formed from superconducting material—the first one in use for n.m.r. studies had a field of 50,000 gauss, and a corresponding operating frequency for hydrogen nuclei of some 200 Mc/sec. There are two advantages in operating at high field strengths; firstly, the intensity of the signal should increase with the applied field, so more sensitive spectrometers can be built, and secondly the chemical shifts increase, the spectrum is more spread out and so more positive analyses for groups become possible. Superconducting magnets are, however, still in their infancy and it will be some time before they become commercially available.

Electron spin resonance spectroscopy imposes far less stringent conditions on the magnetic field than n.m.r., since the spectra show much wider absorption lines and much larger coupling constants (these are measured in gauss for e.s.r. while they are fractions of a milligauss in n.m.r.). Quite a small magnet is usually adequate, but it is invariably an electromagnet since the large sweep fields required to scan a spectrum would rapidly spoil a permanent magnet by hysteresis.

We have mentioned earlier that e.s.r. spectroscopy, as usually carried out at 3000 gauss, falls in the microwave region of the spectrum. In all respects, other than the provision of a magnet, the techniques of e.s.r. are very similar to those of rotational spectroscopy discussed in Chapter 2; the reader is referred to Section 5 of that chapter for brief details of the system.

Nuclear magnetic resonance spectroscopy, on the other hand, is carried on in the radiofrequency region and the techniques used are essentially those of v.h.f. radio. We shall discuss the components very briefly here.

(i) *Source.* This is essentially a carefully tuned crystal oscillator. The frequency generated must be maintained constant to within at least the accuracy of the magnetic field, i.e. at least 1 in 10^8; this is some 0·5 c/sec at 50 Mc/sec, and again is a very stringent requirement. To this end the generating potential of the oscillator is maintained constant to very fine limits and the temperature of the crystal itself is very carefully controlled. Since it is very much easier to produce a source of constant frequency rather than one accurately tunable over a range of frequencies, most spectrometers vary the field strength in order to scan a spectrum.

(ii) *Directing and Focusing the Radiation.* This is particularly simple in n.m.r. spectroscopy, since the "beam" of radiation is carried along coaxial cables from the source to the sample; directing and focusing by this means are trivial. The sample is bathed in radiation emitted from small oscillator or transmitter coils fed from the source and, if absorption spectra are being studied, the loss in energy of the transmitter coil is measured; a sudden loss at a particular field strength shows that resonance absorption has occurred. Emission (or *nuclear induction*) spectra may be studied equally easily by placing a receiver coil (at right angles to the transmitter so that no direct pick-up occurs) round the sample; any emission from the sample is collected by this coil and its output amplified and recorded by normal v.h.f. techniques.

(*iii*) *The Sample.* For nearly all chemical studies this is a liquid, either pure or in solution; solid samples, as discussed earlier, yield rather broad and chemically uninteresting spectra (although much structural information can be obtained) while gas samples normally give absorption or emission signals too weak to be readily detected. The liquid is held in a glass tube some 5 mm in outside diameter and placed between the magnet poles inside the transmitter and receiver coils. It is then spun about its cylindrical axis to improve the apparent field homogeneity as mentioned above.

Bibliography

BERL, W. G. (Ed.): *Physical Methods in Chemical Analysis*, Vol. III, Academic Press, 1956

CARRINGTON, A.: Electron-spin Resonance Spectra of Aromatic Radicals and Radical-ions, *Quart. Rev.*, 1963, Vol. XVII, No. 1, p. 67, The Chemical Society, London

EMSLEY, J. W., J. FEENEY and L. H. SUTCLIFFE: *High Resolution Nuclear Magnetic Resonance Spectroscopy*, Vol. I, Pergamon Press, 1965

POPLE, J. A., W. G. SCHNEIDER and H. J. BERNSTEIN: *High Resolution Nuclear Magnetic Resonance*, McGraw-Hill, 1959

ROBERTS, J. D.: *An Introduction to the Analysis of Spin-Spin Splitting in N.M.R.*, Benjamin, 1961

ROBERTS, J. D.: *Nuclear Magnetic Resonance*, McGraw-Hill, 1959

WALKER, S. and H. STRAW: *Spectroscopy*, Vol. I, Chapman & Hall, 1961

VARIAN ASSOCIATES: *NMR and EPR Spectroscopy*, Pergamon Press, 1960

INDEX